FLEXIBLE MANUFACTURING CELLS AND SYSTEMS

William W. Luggen

PRENTICE HALL, *Englewood Cliffs, New Jersey 07632*

670.427
LUG

Library of Congress Cataloging-in-Publication Data

Luggen, William W.
    Flexible manufacturing cells and systems / William W. Luggen.
        p.   cm.
    Includes bibliographical references.
    ISBN 0-13-321738-8
    1. Flexible manufacturing systems.   I. Title.
    TS155.6.L84   1991
    670.42'7--dc20   89-71022
                                                        CIP

Editorial/production supervision and
  interior design: Shelly Kupperman
Cover design: Wanda Lubelska
Manufacturing buyer: Margaret Rizzi
Cover Photo: Courtesy of General Dynamics, Fort Worth Division.

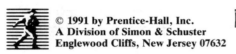

© 1991 by Prentice-Hall, Inc.
A Division of Simon & Schuster
Englewood Cliffs, New Jersey 07632

I 0 6 4 8 5 3   4

The author and publisher of this book have used their best efforts in preparing this
book. In some photographs, safety equipment may have been removed or opened to
more clearly illustrate the product or machinery. For safe operation at all times, the
safety equipment must be in position prior to operation. The author and the publisher
shall not be liable in any event for incidental or consequential damages in connection
with, or arising out of, the use of any machinery or products featured in this book.

Printed in the United States of America
10 9 8 7 6 5 4 3 2 1

ISBN   0-13-321738-8

PRENTICE-HALL INTERNATIONAL (UK) LIMITED, London
PRENTICE-HALL OF AUSTRALIA PTY. LIMITED, Sydney
PRENTICE-HALL CANADA INC., Toronto
PRENTICE-HALL HISPANOAMERICANA, S.A., Mexico
PRENTICE-HALL OF INDIA PRIVATE LIMITED, New Delhi
PRENTICE-HALL OF JAPAN, INC., Tokyo
SIMON & SCHUSTER ASIA PTE. LTD., Singapore
EDITORA PRENTICE-HALL DO BRASIL, LTDA., Rio de Janeiro

20/7/93

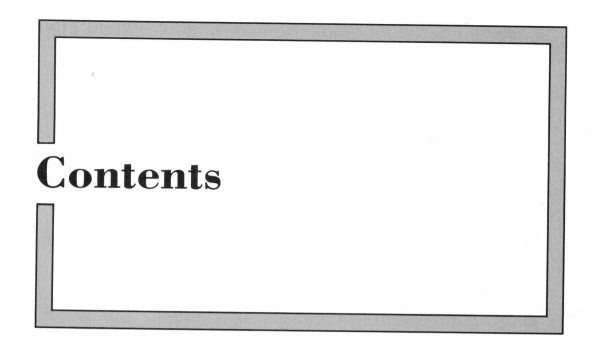

Contents

3   MANUFACTURING CELLS   **40**

Part II: Systems Planning   **63**

4   PHYSICAL PLANNING   **64**

5   HUMAN RESOURCES   **105**

6   QUALITY: MANUFACTURING'S DRIVING FORCE   133

7   JUST-IN-TIME MANUFACTURING   145

8   GROUP TECHNOLOGY   157

Part III: Processing and Quality Assurance Equipment 169

Contents

Part IV: System Support Equipment 249

Preface

Our manufacturing world continues to become more complex brought about principally by global competition, cost and profitability pressures, and rapidly advancing technology. Flexible manufacturing cells and systems represent an avenue of change for manufacturers that helps to bridge the gap between technology, competition, and profitability through a highly specialized and focused approach to manufacturing effectiveness.

The concept of flexible automation evolved through efforts to more efficiently and effectively utilize and control assets, information, and resources in a changing economic climate. Automated cells and systems have grown from concept to reality in light of the installed systems currently in place throughout the United States, Japan, and Europe. And they are projected to grow dramatically over the next several years as manufacturers strive to become more comfortable with the increasing installed base of cells and systems and the dynamic forces of competitiveness, productivity, and profitability.

Although the principal focus of flexible cells and systems in this text is toward the part manufacturing industries, many automated systems, because they are defined differently by their users, have and will continue to be used for process-oriented and assembly applications. It is important for students to recognize the broad implications of flexible automation; it is the complete change in thinking and attitudes of management and labor that really makes it work. Consequently, many of the cell and FMS principles and practices discussed in this text can apply to all types of industries.

Students must maintain a sense of perspective while using this text and recognize that flexible automation is not an end in itself, but a means to an end. Cells and systems represent only a piece of the evolving factory automation puzzle. Just as numerical control is only a small part of an FMS installation, so also is FMS only a small piece of the larger and more complex puzzle called computer-integrated manufacturing (CIM).

Each evolving technology has its place and purpose in the rapidly changing world of manufacturing automation. It is only through an individual's understanding and recognition of this total perspective that integration of the various pieces will ultimately come together to form CIM and the computer-integrated enterprise.

This text has been written to be more than simply a textbook. It was written to be a comprehensive, user-friendly reference guide that will benefit students, technicians, and engineering professionals with descriptive and easy to understand explanations throughout their careers. Information in the text covers topics from general FMS description and considerations to factors relating to systems planning, human resources, quality control, and group technology. Detailed but easy to read and understand explanations are provided for part processing and material-handling equipment, tooling, fixturing, communication networks, robots, turning centers, machining centers, and computer hardware and software. Additional practical and concise information is provided for FMS implementation, installation, tool management, maintenance concerns, and training issues.

Flexible manufacturing cells and systems is an application-oriented text written to supply both system user and system supplier points of view. It is arranged in six primary parts and nineteen chapters with eighty informative sub-topics and over 300 photographs and illustrations. Each chapter begins with an opening introduction and specific learning objectives and ends with key summary points and review questions. Supplemental accompaniments to the text include appendixes for avoiding problems and pitfalls, general safety rules, and EIA and AIA national codes, along with a descriptive glossary.

Topics can be taught sequentially because they are arranged in the table of contents in a progressive concept-to-implementation manner. Or individual topics can be taught out of order to suit the course requirements or the instructor's approach. Primary prerequisite courses prior to using this text would consist of modern manufacturing processes, CNC systems, and computer fundamentals.

This text is intended to be used for any course in universities, community colleges, and adult retraining programs where general education about flexible automation is required–what it is, what it does, what and who is involved, and why it is important. These and many other questions are answered in great detail in this text in a friendly and nonintimidating fashion.

My sincere hope is that you will find the text interesting, informative, and easy to use and that in some way I will have touched each of you, the students, engineers, technicians, and decision makers of tomorrow.

ACKNOWLEDGMENTS

The author and publisher recognize with appreciation the technical assistance and photographs that the following companies provided:

Allen-Bradley Co.
Automated Finishing, Inc.
Cincinnati Industrial Machinery, Inc.
Cincinnati Milacron, Inc.
Cushman Industries
DeVlieg Sundstrand, Inc.
Eaton-Kenway, Inc.
General Dynamics, Fort Worth Division
GMFanuc Robotics Corp.
ICL Manufacturing, Inc.
Interlake Material Handling Division
International Business Machines Corporation
ITW Woodworth, Inc.
Jergens Power Clamping, Inc.
Jervis B. Webb Co.
Jorgensen Conveyors, Inc.

Kearney and Trecker Corporation
Kennametal, Inc.
K. N. Aronson, Inc.
LTV Aircraft Products Group
Maho Machine Tool Co.
Mazak Corporation
Mid-State Machine Products, Inc.
Numeridex, Inc.
Olafsson Co.
Qu-Co Modular Fixturing, A Te-Co Co.
Remington Arms Co.
Renishaw, Inc.
Robert Bosch Corporation—Surf Tran Division
Sheffield Measurement Division
Westinghouse Automation Division

For their critical comments and recommendations, I am indebted to the following reviewers: Jeff Lawrence, Remington Arms Company; D. Peter Jawahar, Western Kentucky University; Robert Romano, Cincinnati Technical College; John Strenge, University of Maryland; and P. Tregoning Southern Illinois University.

William W. Luggen

Man is a user of tools. Those who recognize the tools of tomorrow and learn how to use them today, assure themselves of a place in tomorrow's prosperity.

Anonymous

Part I

Evolution of Manufacturing Systems

Manufacturing industries are under great pressure caused by the rising costs of energy, materials, labor, capital, and intensifying worldwide competition. While these trends will remain for a long time, the problems facing manufacturing today run much deeper. In many cases they stem from the very nature of the manufacturing process itself.

Flexible manufacturing systems (FMS) (Figures P1-1 and P1-2) are regarded as one of the most efficient methods to use in reducing or eliminating problems in the manufacturing process. FMS is more than a technical solution; it is a business-driven solution leading to improved profitability through reduced lead times and inventory levels, rapid response to market changes, lower staffing levels, and improved manufacturing effectiveness through increased operational flexibility, predictability, and control.

FMS provides new hope to the manufacturing industries and will continue to be looked on not as an end in itself, but as a dynamic, evolving process to keep the manufacturing world alive and well in the years ahead.

Figure P1-1　Full-scale flexible manufacturing system (FMS). (Courtesy of Cincinnati Milacron)

Figure P1-2   Flexible manufacturing systems are regarded as one of the most efficient means of reducing problems in the manufacturing process. (Courtesy of General Dynamics, Fort Worth Division)

FMS: Definition and Description

OBJECTIVES

After studying this unit you will be able to:

- □ Broadly describe FMS, what it consists of, and what it does.
- □ Name some common day to day disturbances in a manufacturing operation.
- □ List some primary goals and objectives of FMS.

INTRODUCTION

Definitions and descriptions abound for FMS. In many cases it depends on the user's rather than the equipment manufacturer's point of view. FMS technology, originating in Europe, has evolved over the last two decades primarily to meet the requirements of the mid-volume, mid-variety world of part manufacturing. This focus has occurred because part manufacturing is one of the most costly and unprofitable areas in industry. However, FMS is not limited to part manufacturing operations. Depending on the user's definition of FMS and the type of industry, FMS has been and will continue to be utilized in process- and assembly-related industries as well as part manufacturing. Many of the principles and practices apply regardless of industry type, business objectives, or definition of FMS.

FMS changes people's attitudes in just the same way it needs a changed attitude to introduce it. FMS brings flexibility and responsiveness to the manufacturing floor—when a part is required by the market and not when production deems it so. FMS provides additional flexibility to adjust to changing market conditions and product mixes without further investment. FMS releases dormant and unprofitable machining capacity, particularly during second and third shifts, often regarded as manufacturing's "unsocial hours." It is a major steppingstone to unmanned manufacturing, the automated factory, and the promise of tomorrow.

EVOLUTION

The concept of flexible manufacturing systems was born in London in the 1960s when David Williamson, a research and development engineer, came up with both the name and the concept. At the time he was thinking in terms of a flexible machining system, and it was in a machine shop that the first FMS was installed. His concept was called System 24 because it was scheduled to operate for 24 hours a day under the control of a computer, but otherwise unmanned on the 16-hour night shift. This simple concept of decentralized computer control of machine tools, combined with the idea of using machine tools for 24 hours per day (16 unmanned on night shift), was the beginning of FMSs.

Williamson planned to use NC (numerically controlled) machines to work out a series of machining operations on a wide range of detail parts. Workpieces would be loaded manually on pallets, which would then be delivered to the machines and loaded automatically when needed. Each machine would be equipped with a magazine from which tools could be selected systematically to perform a variety of different operations. Included in this overall process were systems for removing chips and cleaning workpieces. This system combined the versatility of computer-controlled machines with very low manning levels.

With the growth in computer-controlled equipment and broader applications developing from metal forming to assembly, the concept of "flexible machining systems" was broadened to become what is known today as "flexible manufacturing systems," or FMS.

As the first FMS systems were installed in Europe, they followed Williamson's concept, and users quickly discovered that the principles would be ideal for the manufacture of low-volume, high-variety products. The addition of refinements to an FMS to detect and compensate for tool wear were then added to further aid unattended FMS operations. These first FMSs on the market had dual computers: DNC (direct numerical control) for cell control functions and a separate computer to monitor the traffic and management information systems.

Since the 1970s there has been an explosion in system controls and operational enhancements. The programmable controller appeared in the late 1970s, and the personal computer emerged utilizing distributed logic control with many levels of intelligent decision-making capabilities.

Thus, through a conceptual idea originating with David Williamson, it became possible to machine a wide variety of workpieces on few machines with low manning levels productively, reliably, and predictably; this is what FMS is all about. In almost any manufacturing industry, FMS will pay dividends as long as it is applied in its broad sense, and not just to define a machining system.

The FMS has evolved rapidly and will continue to evolve because technology continues to evolve, global competition intensifies, and the concept of flexible manufacturing gains wider acceptance. The growth of flexible manufacturing is projected to increase steadily in the years ahead.

In 1984, 56 percent of all FMSs were used for manufacturing machinery and 41 percent for manufacturing transportation components. Construction and material-handling industries will comprise around 12 percent of the user market in the early 1990s as their adoption of FMS increases.

Flexible automation is presently feasible for a few machining operations that account for a fraction of the total manufacturing process. However, development efforts continue to expand the FMS's capabilities in the areas of improved diagnostics and sensors, high speed, noncontact, on-line inspection, multifunction or quick spindle head changing machine tools, and extending flexible automation to include forming, heat treating, and assembly. This is why FMS continues to grow and prosper. It feeds on technology evolving and expanding as technology itself evolves and expands.

WHAT IS FMS?

Definitions of FMS, or Flexible Manufacturing Systems, are plentiful and in many respects are dependent on the ultimate user's point of view as to what the FMS consists of and how it will be used. However, the following represent a collection of FMS definitions, some traceable and some not traceable to their originating source.

1. United States Government: A series of automatic machine tools or items of fabrication equipment linked together with an automatic material handling system, a common hierarchial digital preprogrammed computer control, and provision for random fabrication of parts or assemblies that fall within predetermined families.

2. Kearney and Trecker: A FMS is a group of NC machine tools that can randomly process a group of parts, having automated material handling and central computer control to dynamically balance resource utilization so that the system can adapt automatically to changes in parts production, mixes, and levels of output.

3. FMS is a randomly loaded automated system based on group technology manufacturing linking integrated computer control and a group of machines to automatically produce and handle (move) parts for continuous serial processing.

4. FMS combines microelectronics and mechanical engineering to bring the economics of scale to batch work. A central on-line computer controls the machine tools, other workstations, and the transfer of components and tool-

ing. The computer also provides monitoring and information control. This combination of flexibility and overall control makes possible the production of a wide range of products in small numbers.

5. A process under control to produce varieties of components or products within its stated capability and to a predetermined schedule.

6. A technology which will help achieve leaner factories with better response times, lower unit costs, and higher quality under an improved level of management and capital control.

Regardless of how broadly or narrowly FMS is defined, several key items emerge as critical to a general definition of FMS, and repeat themselves through a cross-section of standard definitions. Words like *NC machine tools, automatic material handling system, central computer controlled, randomly loaded, linked together* and *flexible* all serve to help define a very general description and definition of FMS.

Flexible manufacturing systems are based on modular part producing machinery—machine tools, or injection molding machines, for example, and a wide variety of ancillary support equipment, linked and integrated together under central computer control to produce a variety of components in random order.

Basically, an FMS is made up of hardware and software elements. Hardware elements are visible and tangible such as CNC machine tools, pallet queuing carrousels (part parking lots), material handling equipment (robots or automatic guided vehicles), central chip removal and coolant systems, tooling systems, coordinate measuring machines (CMMs), part cleaning stations, and computer hardware equipment. Software elements are invisible and intangible such as NC programs, traffic management software, tooling information, CMM program work-order files, and sophisticated FMS software. A typical FMS layout and its major identifiable components can be seen in Figure 1-1.

A true FMS can handle a wide variety of dissimilar parts, producing them one at a time, in any order, as needed (very few so-called FMSs meet this strict definition). To adapt efficiently in this mode, an FMS must have several types of flexibility. It needs the flexibility to adapt to varying volume requirements and changing part mixes, to accept new parts, and to accommodate design and engineering modifications. FMS also requires the flexibility to cope with unforseen and unpredictable disturbances such as machine downtime problems or last minute schedule changes; and the ability to grow with the times through system expansion and configuration, improvements, and alterations. These types of flexibility are made possible through computers and appropriate FMS software. It is the sophisticated FMS software that actually "drives" the total system and provides the decision criteria to anticipate resource needs, schedule production, and respond to planned and unplanned real-time activities.

Because FMS is a business—driven solution to mid-volume, mid-variety manufacturers, it offers the opportunity of predictable control to the manufacturing process. FMS is a process under total control. A full FMS installation is one in which a process is put under total computer control to produce a variety of

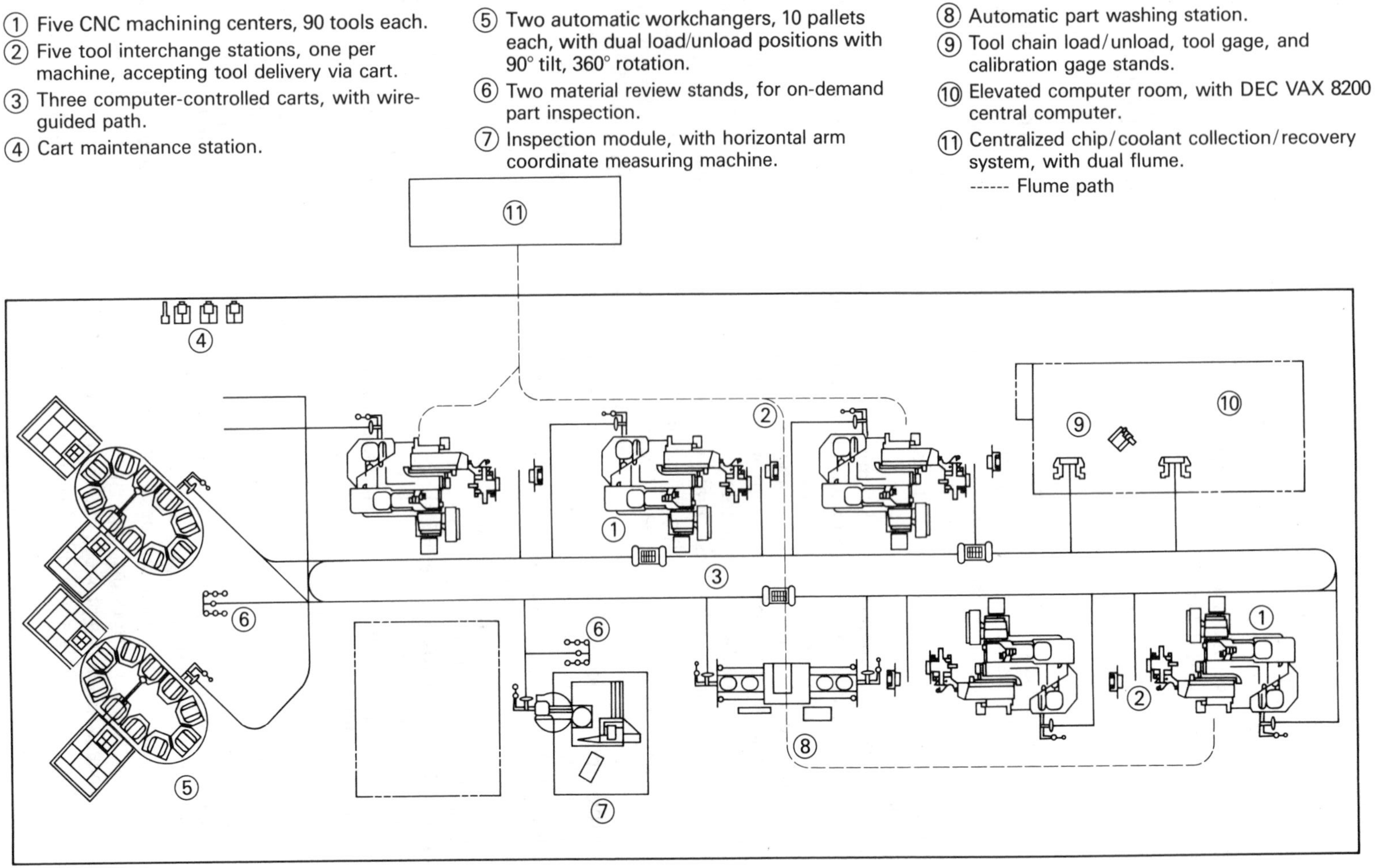

① Five CNC machining centers, 90 tools each.
② Five tool interchange stations, one per machine, accepting tool delivery via cart.
③ Three computer-controlled carts, with wire-guided path.
④ Cart maintenance station.
⑤ Two automatic workchangers, 10 pallets each, with dual load/unload positions with 90° tilt, 360° rotation.
⑥ Two material review stands, for on-demand part inspection.
⑦ Inspection module, with horizontal arm coordinate measuring machine.
⑧ Automatic part washing station.
⑨ Tool chain load/unload, tool gage, and calibration gage stands.
⑩ Elevated computer room, with DEC VAX 8200 central computer.
⑪ Centralized chip/coolant collection/recovery system, with dual flume.
----- Flume path

Figure 1-1 Typical FMS and its major identifiable components. (Courtesy of Cincinnati Milacron)

products with the system's defined capability and with a pre-determined schedule.

In the long range, FMS is the natural partner for CAD (Computer Aided Manufacturing) and CIM (Computer Integrated Manufacturing) which ultimately all serve to bring a product from design to reality by the most efficient and cost-effective means.

In an FMS installation, the moment-by-moment functions, actions, and decisions are inherent within the system—operating completely without (or with very little) human intervention. These moment-by-moment activities involve not only material handling, but also inspection, part washing, tool storage, fixturing, and warehousing, in addition to downloading of NC programs and other normal machine functions.

Depending on a company's specific manufacturing needs, an FMS may or may not be the answer. The graph in Figure 1-2 illustrates the range of application solutions available for a given set of workpiece volume a variety requirements. An FMS is set apart from any other kind of manufacturing system, such as a transfer line used in high volume automotive applications, because of its ability to accept parts or components in varying quantities, in random order. Thus, an FMS can be designed to process any product, in any volume, in any order, within the family of components designed for the system.

By definition, an FMS can simultaneously process a variety of workpieces, using tooling and fixturing made available at the right machine, at the right time, and in the right sequence. The FMS computer functions to identify these needs and

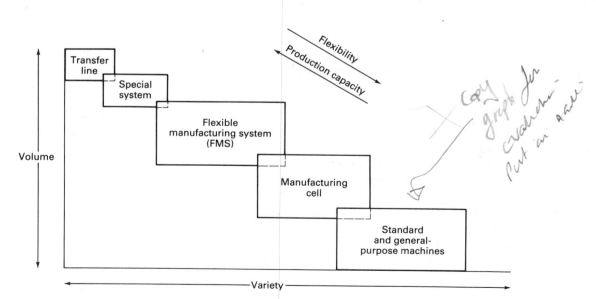

Figure 1-2   General range of application solutions based on a given set of workpiece volume and variety requirements. (Courtesy of Kearney and Trecker Corp.)

allocates resources in the form of tooling, fixtures, material movements, and NC and inspection programs in order to fulfill predetermined work order requirements.

Is there an optimum size of FMS? At the present time the answer is no; size depends on user needs and resources. The number of NC machines in a system, for example, can be as low as one or two. This can provide a starting point for those who wish to take advantage of FMS in a step-by-step or phased-in approach.

Generally, the number of processing machines or machine tools is three to ten. Thus, the cost and effort involved for planning, project engineering, installing, implementing, and managing an FMS is high.

NEED FOR FMS

The key objective in manufacturing is to get the right raw materials or parts to the right machines at the right time. Too much or too soon creates backed up excess in-process inventory. Too little or too late causes delayed work schedules and idle machines. The result in many cases is a poor use of capital, in the form of excess in-process inventory and/or underutilization of equipment.

In any single calendar year, there are 8760 hours available to the manufacturing operation, as can be seen in Figure 1-3. Statistics have shown that about 44 percent of the total time available is lost due to incomplete use of second and third shifts. The skilled, experienced people required to operate and set up machines are either not available or disinterested in working "unsocial" hours, and the problem is going to get worse. The long-term trend is firmly established that a declining percentage of people entering the work force will choose careers in manufacturing.

Thirty-four percent of the total time is lost due to vacations and holidays. Twelve percent is lost while machines are being set up for the next operation or parts are being loaded or unloaded. About 4 percent of the time is lost due to process difficulties or unforeseen material, tooling, or quality-control problems.

This leaves only 6 percent of the total time for actual production. The batch manufacturer's capital investment for equipment and facilities is working, trying to pay for itself, less than one hour in seventeen.

Similar studies indicate that in a typical manufacturing operation a part moving through a metal-cutting operation would be on an individual machine tool only 5 percent of its total time in manufacturing, as depicted in Figure 1-4. And, when a part is on a particular metal-cutting machine tool, only 1.5 to 2 percent of the part's total manufacturing time is a cutter in the work, actually performing work and adding value. The other 95 percent of the time the part is either moving through the shop or waiting in queue for the next operation.

These examples indicate the underutilization of equipment and gross inefficiencies existing in a vast majority of manufacturing industries. Many of these inefficiencies are common day to day disturbances within the overall manufacturing process consisting of:

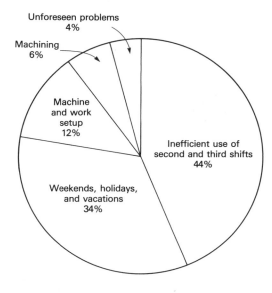

Unforeseen problems
4%

Machining
6%

Machine
and work
setup
12%

Inefficient use of
second and third shifts
44%

Weekends, holidays,
and vacations
34%

**Figure 1-3   Breakdown of 8760
available hours in a calendar year
to a manufacturing operation.**

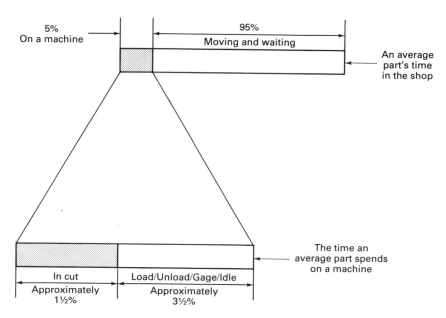

5%
On a machine

95%
Moving and waiting

An average
part's time
in the shop

The time an
average part spends
on a machine

In cut
Approximately
1½%

Load/Unload/Gage/Idle
Approximately
3½%

Figure 1-4   Breakdown of the time spent by an average part in the shop.

1. Priority (scheduling) changes
2. Engineering design changes
3. Tooling difficulties
4. Machine breakdowns
5. Processing problems
6. Lost, misplaced, and scrapped parts
7. Vendor lateness

What is needed in today's competitive environment, regardless of what products a particular company makes, is the capability to effectively manage and control the day to day disturbances while meeting customer requirements. This implies that:

1. There should be minimum delay between order placement and order delivery.
2. Quality and reliability should be high.
3. Operating costs should be predictable and under control.
4. Replacement parts should be available and accessible on a quick turnaround basis.

FMS provides a means to manage and control the uncontrollable disturbances while meeting customer demands and requirements.

The principal objectives of FMS are:

1. Improve operational control through:
   a. Reduction in the number of uncontrollable variables
   b. Providing tools to recognize and react quickly to deviations in the manufacturing plan
   c. Reducing dependence on human communication
2. Reduce direct labor through:
   a. Removing operators from the machining site (their responsibilities and activities can be broadened)
   b. Eliminating dependence on highly skilled machinists (their manufacturing skills can be better utilized in manufacturing engineering functions)
   c. Providing a catalyst to introduce and support unattended or lightly attended machine operation
3. Improve short-run responsiveness consisting of:
   a. Engineering changes
   b. Processing changes
   c. Machine downtime or unavailability
   d. Cutting tool failure
   e. Late material delivery
4. Improve long-run accommodations through quicker and easier assimilation of:

a. Changing product volumes
b. New product additions and introductions
c. Different part mixes

5. Increase machine utilization by:
a. Eliminating machine setup
b. Utilizing automated features to replace manual intervention
c. Providing quick transfer devices to keep machines in the cutting cycle

6. Reduce inventory by:
a. Reducing lot sizes
b. Improving inventory turnovers
c. Providing the planning tools for just-in-time manufacturing

These objectives are broad, far-reaching and not easily attainable. However, work-in-process costs alone have tripled in the last ten years, and increased pressures for reduced costs, increased quality, and improved delivery continue unabated. FMS provides a well-centralized and coordinated means of shortening the manufacturing pipeline and obtaining these goals through a comprehensive and focused approach.

SUMMARY

1. Flexible manufacturing systems are regarded as one of the most efficient methods to employ in reducing or eliminating problems in manufacturing industries.

2. FMS originated in London, England, in the 1960s when David Williamson came up with a flexible machining system called System 24 to operate unmanned 24 hours a day under computer control. Broader applications developed and continue to be developed in the areas of injection molding, metal forming and fabricating, and assembly, thereby broadening the name to flexible manufacturing systems.

3. Definitions of FMS vary depending on industry type and the user's point of view.

4. Many FMS principles and practices apply regardless of industry type, business objectives, or line of product.

5. FMS brings flexibility and responsiveness to the manufacturing floor.

6. Since the 1970s there has been explosive growth in system controls and operational enhancements, which has allowed FMS to grow, develop, and gain wider acceptance.

7. FMS enables manufacturers to machine a wide variety of workpieces on few machines with low staffing levels, productively, reliably, and predictably.

8. FMS is made up of hardware elements (machine tools, movable pallets, material-handling equipment, coordinate measuring machines, computer hardware equipment, and the like) and software elements (NC programs, inspection programs, work-order files, and FMS software). The sophisticated FMS software is what actually drives the system.

9. A true FMS can handle a wide variety of different parts, producing them one at a time in random order.

10. FMS is not an end in itself, but a means to an end and the natural partner to integrate to existing CAD/CAM systems and progress toward CIM.

11. Machine tools in many manufacturing industries are woefully under-utilized due to equipment not being used on second and third shifts, a decreasing availability of skilled personnel, and day to day disturbances.

12. FMS shortens the manufacturing process through improved operational control, round-the-clock availability of automated equipment, increased machine utilization and responsiveness, and reduction of human intervention.

REVIEW QUESTIONS

1. FMS originated in the United States and the technology was transferred to Europe. True or False?
2. Many FMS principles and practices are applicable to productivity improvements regardless of industry and product type. True or False?
3. Williamson's first FMS was called _____.
4. The definition of FMS:
   (a) Applies only to specific industries
   (b) Must always include metal-cutting machine tools
   (c) Can vary depending on the user's point of view, what the FMS consists of, and how it will be used
   (d) Only relates to systems automatically transferring material
5. Briefly explain why many manufacturer's equipment is busy cutting metal only 6 percent of the time.
6. An FMS is similar to an automotive transfer line because it can only accept parts within a tightly defined part family and only in high volumes. True or False?
7. The key objective in manufacturing is to get the right parts to the right machines at the right time. True or False?
8. The majority of manufacturing time is lost due to:
   (a) Vacations and holidays
   (b) Incomplete use of second and third shifts
   (c) Wrong processes, scrapped parts, and lateness
   (d) Only (a) and (b)
   (e) All of the above
9. FMS will reduce inventory by increasing inventory turnover. True or False?
10. Which of the following is *not* a typical day to day disturbance within the overall manufacturing process?

(a) Workpiece problems
(b) Processing problems
(c) Tooling difficulties
(d) Power failures
(e) Engineering changes

11. An important objective of FMS is to increase machine utilization by using automated features to replace manual intervention. True or False?

2

General FMS Considerations

OBJECTIVES

After studying this unit, you will be able to:

' □ Identify some common problems associated with conventional manufacturing.

□ Name and describe key FMS benefits.

□ Discuss the principal factors related to a winning FMS justification plan.

□ Explain the importance of management commitment to a successful FMS.

INTRODUCTION

Many manufacturing industries are currently dedicated to manual and conventional production methods or high-speed fixed automation—by their very nature not very flexible or responsive. And many are ill suited to accommodate faster product and process changes in an increasingly globalized and competitive marketplace.

Flexible manufacturing affords users the opportunity to react quickly to changing product types, mixes, and volumes while providing increased utilization and predictable control over hard assets. Although FMS provides users with many benefits, they are not easy to justify. Limitations and alternatives must be weighed and compared to determine if FMS is the best or even the right approach to productivity and profitability improvements. The once traditional accounting and cost justification practices have become outdated and have lost their applicability to many factory automation programs and projects. The rules for staying competitive have changed. The measurements must also change.

CONVENTIONAL APPROACHES TO MANUFACTURING

Conventional approaches to manufacturing have generally centered around machines layed out in logical arrangements in a manufacturing facility. These machine layouts are usually classified by:

1. Function
2. Line
3. Cell

A machine layout by function is the arrangement of similar machines by department, as seen in Figure 2-1. This is the typical mill department, drill department, grind department, and so on. Machines organized by function will typically perform the same function, and the location of these departments relative to each

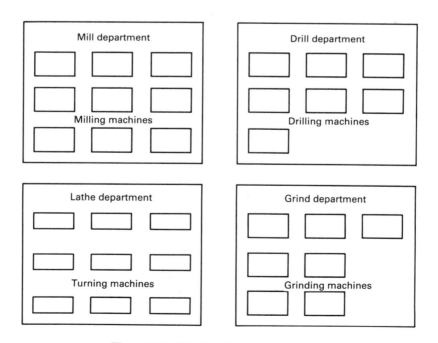

Figure 2-1   Machine layout by function.

other is normally arranged so as to minimize interdepartmental material handling. Workpieces produced in functional layout departments and factories are generally manufactured in small batches up to fifty pieces. Functional machine arrangements can process a great variety of parts and are characteristic of many manufacturing facilities, including the vast multitude of job shops.

A line or flow layout is the arrangement of machines in the part processing order or sequence required. A typical part consisting of milling, drilling, grinding, and inspection operations for example, would require a line arrangement consisting of a mill, a drill, a grinder, and an inspection operation. In a line layout, machines are located next to each other in the order in which they will be used, as can be seen in Figure 2-2. A transfer line is an example of a line layout. Parts progressively move from one machine to another in a line or flow layout by means of a roller conveyor or through manual material handling. Typically, one or very few different parts are produced on a line or flow type of layout, as all parts processed require the same processing sequence of operations. All machining is usually performed in one department, thereby minimizing interdepartment material handling. Intradepartment material handling is also minimized because successive operations are performed on machines next to one another. Line layouts are not as widely used throughout industry as the traditional functional layout.

The cell layout is a combination of both the function and line layouts. It combines the efficiencies of both layouts into a single multifunctional unit. Sometimes referred to as a group technology cell, each individual cell or department is comprised of different machines that may not be identical or even similar, as can be

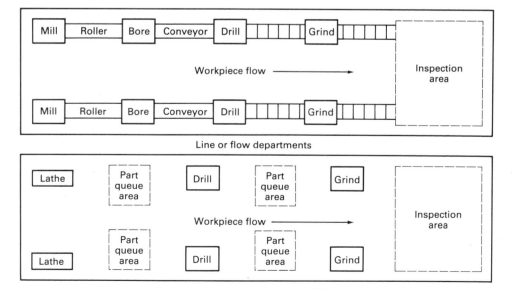

Figure 2-2   Line or flow machine layout.

seen in Figure 2-3. Each cell is essentially a factory within a factory, and parts are grouped or arranged into families requiring the same type of processes, regardless of processing order. Cellular layouts are highly advantageous over both function and line machine layouts because they can eliminate complex material flow patterns (Figure 2-4) and consolidate material movement from machine to machine within the cell (Figure 2-5). Complex material flow patterns create extensive part move and queue time, result in lost or misplaced parts, and contribute to damaged parts due to excessive movement. Consequently, their elimination is highly desirable. Additionally, cell arrangements can accommodate a variety of dissimilar parts within the designated workpiece family to be processed through the cell. Part family cells essentially tradeoff interdepartment material handling for intradepartment material handling.

   Although each type of layout brings certain advantages and disadvantages to the manufacturing floor, the vast inefficiencies existing in both process and discrete industries coupled with intensified global competition are forcing many manufacturers to rethink and reevaluate the way they run their businesses. While the bottom line for all manufacturing is still the production of a quality product at a reasonable price with acceptable delivery, it is no longer sufficient to depend on product innovation, marketing tactics, and salesmanship to maintain competitive leadership. Manufacturing has emerged as the competitive battleground, and sheer survival hinges on the ability to compete as a cost-effective manufacturer in an environment of escalating competitive pressure.

   What are the problems facing manufacturing industries today? What has

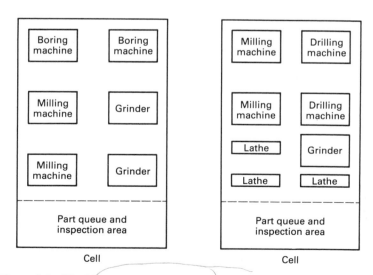

Figure 2-3   Machine layout by cell based on part families to be processed in each cell.

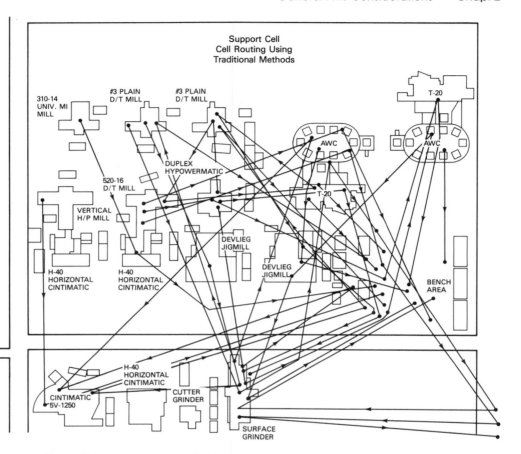

Figure 2-4 Complex material flow pattern indicating many interdepartmental moves and great distances a part must travel to comlete processing. (Courtesy of Cincinnati Milacron)

changed to allow these inefficiencies to coexist? To begin with, consumer demand and dissatisfaction with mediocrity have intensified the scale of the problem due to the multitude of product variations that now must be offered. The complexity and exacting nature of many products require more rigorous planning, engineering, distribution, and manufacturing capabilities. Today's competitive climate means higher management expectations in terms of profitability, less time to react to changes in the marketplace, less forgiving consequences as a result of bad decisions, and managing with less cushion.

Problems associated with conventional manufacturing include both external and internal pressures and inefficiencies. External pressures result from:

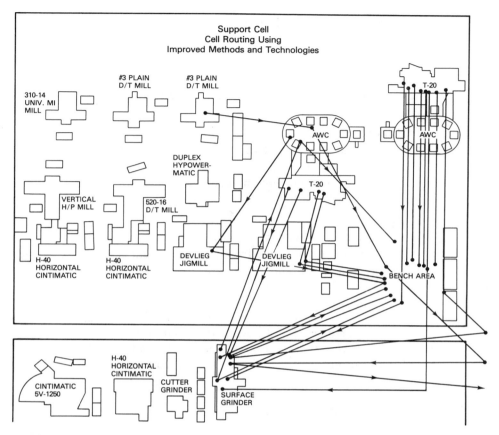

Figure 2-5 Intradepartmental material flow within a cell minimizes interdepartment material moves and queue time. (Courtesy of Cincinnati Milacron)

- ☐ Technological advancements (let's do something so we don't technically fall behind)
- ☐ Increased cost, quality, and delivery pressure as a result of intensifying worldwide competition
- ☐ Fluctuating exchange rates (closer scrutiny of make versus buy decisions)
- ☐ Uncertainty and instability of economic conditions
- ☐ Declining percentage of individuals choosing careers in manufacturing

Internal problems and inefficiencies result from:

- ☐ High levels of work-in-process inventories

☐ Complex material flow patterns
☐ Extremely long lead times
☐ Increasing product complexities
☐ Excessive material handling and damaged parts
☐ Complex scheduling and machine capacity loading
☐ Low capital asset utilization
☐ "Shop floor" engineering changes
☐ Bottlenecked machine groups as a result of multiple parts competing for the same work center
☐ Aging capital equipment and inadequate allocation of replacement funds
☐ Excessive expediting as a result of front-loaded lateness
☐ Inadequate or unavailable tooling
☐ Excessive move, queue, and part setup time
☐ Misplaced parts resulting from part movement to unofficial queue areas
☐ Out of control scrap and rework costs

Manufacturing industries continue to cope with these and other problems in an era of faster change but slower growth. Many manufacturers see a comprehensive, well-planned, well-orchestrated manufacturing automation program as the only way to battle both internal and external problems, regain control of operations, and return to improved levels of productivity and profitability.

BENEFITS AND LIMITATIONS

Manufacturers who invest in flexible manufacturing cells and systems are investing in a result. Flexible manufacturing is the beginning of a new industrial revolution that will conceivably lead the manufacturing industry to levels of automation taken for granted today in the process-related industries. FMS offers to manufacturers a concept that not only will boost productivity, but an entire strategy that also will change the way in which companies operate from internal purchasing and ordering procedures to distribution and marketing. Additionally, FMS represents an unprecedented opportunity to slash dramatically the hidden high costs of manufacturing work-in-process inventory (Figure 2-6) and overhead such as indirect labor. Increasing surges in material and overhead costs, for example, can account for 50 to 90 percent of total manufacturing costs; consequently, any reductions here become quite impressive.

The four principal benefits derived from implementing FMS are:

1. Inventory reduction of 60 to 80 percent
2. Direct labor savings of 30 to 50 percent
3. Increased asset utilization approaching 80 to 90 percent
4. Floor space reduction of 40 to 50 percent

Inventory reduction is a key benefit because parts do not sit around 95 percent of the time waiting to be used. FMS provides the capability for increased part throughput, thus reducing the opportunity for parts to sit around as work in process and finished inventory.

FMS implies low staffing levels, leading to the employment of very few people in direct manufacturing jobs, such as operating machines and assembly. Machinists may know their machines very well and be flexible enough to produce a variety of parts in batches of one and higher. However, they need time to set up their machines, select tools, prepare special fixtures, and so on. Nevertheless, a machinist is a good example of flexibility with poor productivity and low machine utilization. The whole key to FMS is that it offers flexibility and unattended or lightly attended operation as setups and workpieces are fixtured and made off-line while machine tools are cutting metal.

Asset utilization is further increased because equipment can operate lightly staffed for three shifts a day, seven days per week, depending on certain operating conditions and correct equipment balance. The computer-controlled, automated aspects of FMS give the prospect of operation for 24 hours per day. This offers the immediate potential for a massive jump in utilization, which in turn means that floor space can be reduced and thus the actual size and cost of a new plant can be much smaller, perhaps even one-third the size of a conventional plant. Additionally, the space required for work in process and finished inventory further reduces floor space and plant size requirements, because inventory itself is considerably decreased and responsiveness to changes in product mix and volume is increased.

Figure 2-6 Work in process (WIP) inventory is a high-cost item that FMS can help to minimize. (Courtesy of Cincinnati Milacron)

Additional benefits of FMS include the following:

1. Ample capacity in the material movement system and adequate storage for jobs in process to minimize delays waiting for material movement
2. Astute layout of workstations within the system, which minimizes material movement distances and times
3. Processing flexibility to accommodate jobs with different routing and machining requirements
4. Central storage of work in process to reduce the likelihood of bottlenecking certain machines or machine groups
5. Computer-controlled release of jobs to the system so that both bottlenecking and starving of machines can be prevented and the entire workload balanced
6. Monitoring of machine breakdowns and other disturbances in order to divert parts either out of the system or around a particular machine problem without human intervention

Also included would be:

1. Reduced setup time
2. Predictable delivery times
3. Consistent accuracy and high product quality
4. A route and a link to CAD/CAM and CIM (this paves the way to the integrated factory)
5. Reduced lead times
6. Reduced tooling costs
7. Increased profitability
8. Improved manufacturing control
9. Simplified fixture design
10. Reduced overhead costs

FMS further helps to remove some bad manufacturing practices, such as:

1. Damage to parts incurred as a result of ''hurry-up handling'' between the various setups and machining processes. Also, the physical movement of parts from location to location-causes jarring, bumping, and falling, all of which can damage value-added workpieces.
2. Multiple part inspections and the multiple handlings required to support the various inspections. Inspection stations also require queues for parts awaiting rework, along with bins or flats of scrapped parts.
3. Excess just-in-case raw material held in stock for potential scrapped and late parts.
4. Proliferated tool inventories bought because purchasing received ''quantity discounts.''

5. Unaccounted for or lost parts. With FMS it is much harder to lose : system knows where each part is because parts are made only when they a.~ needed.

6. Complex material flow patterns complicating and lengthening part movement from department to department.

As can be seen, the reasons for implementing FMS quickly multiply, many of which derive themselves from the more pervasive benefits of inventory reduction, increased machine utilization, and improved part throughput. In fact, most firms that have installed FMS have done so on the basis of benefits derived from reduced inventory, increased part throughput, predictable quality, and the need to quickly respond to assembly requirements. Installing FMS strictly on the basis of reducing direct labor is a mistake, as direct labor costs account for only about 10 to 15 percent of most products (Figure 2-7). Direct labor reductions occur more as a matter of default through maximized system effectiveness than intentionally by design. FMS induces the discipline of asset utilization and looking at the throughput of the product, rather than eliminating people and slicing 2 minutes from a machining operation.

FMS is not without its inherent limitations. Many of these limitations originate from unrealistic expectations as to what FMS is and what it can do and misunderstanding about the need for flexibility in manufacturing.

Flexibility in manufacturing is an umbrella term for a host of different concepts. Flexibility can be broadly defined as the ability to respond effectively to changing circumstances. However, what constitutes "flexibility" and "changing circumstances" may vary considerably among manufacturers.

Flexibility to some manufacturers means convertibility—being able to convert from manufacturing one product type, family, and/or volume to another within the manufacturer's predetermined time, and not as rapidly as FMS advocates. Thus, convertibility may be the "real" flexibility the manufacturer requires, and it may be done by more astute upgrading or altering of existing resources rather than by the purchase of an FMS.

Flexibility can mean future cost avoidance. This type of flexibility would be common among automotive manufacturers, where high part volumes are common but future changes in market demand are expected and anticipated. Traditional automotive manufacturing employed the use of "hard" or "dedicated" transfer lines. When market demand diminished or changed, a several million dollar transfer line was either idled, underutilized, or discarded. Purchases of flexible transfer lines are now more prevalent among high-volume manufacturers, with the primary thrust of flexibility being future cost avoidance. This helps ensure that future products "fit the system," as opposed to the flexibility of instantaneous changeover for new-model introductions.

Flexibility can also be very narrowly defined and mean predictability among select manufacturers. The predictability aspect of flexibility, for example, can mean the use of lightly attended or unattended one-machine cells where part variation is low (perhaps only two or three different uncomplicated, inexpensive

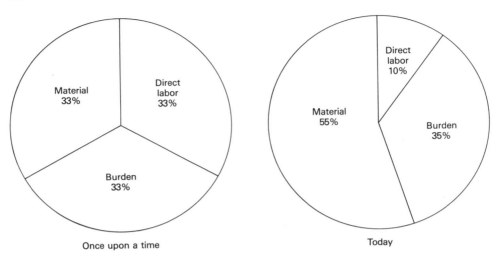

Figure 2-7 Over the years, material costs have increased while direct labor costs have decreased.

parts), but maintaining dependable volume and delivery schedules is critical. This type of cell to some aerospace subcontractors, for example, is a flexible manufacturing system. Here the predictability of manufacture must be precisely "on the money," and dissimilarity varies only within a limited and rather simplistic part spectrum.

Generally, flexibility refers to:

1. *Variety of mix:* the combination of different parts the system can make at a time and the various subsets of part types that can be made simultaneously
2. *Adaptability to design, production, or routing changes:* this refers to ease of accommodating engineering changes, expansion of the total universe of parts producible on the system, and the variety of routes or machines that can process the same part type
3. *Machine changeover:* the ease with which a machine within the system can automatically change from making one part type to another

The essence of this discussion is that flexibility means different things to different manufacturers, and flexibility requirements in manufacturing do not always translate into FMS. The need for FMS is not as important as the overall need for flexibility in manufacturing and in organization within the manufacturer's frame of reference.

Many flexible manufacturing systems have failed or severely faltered because technology was sold for technology's sake, and because technology and its associated buzz words leaped ahead of good business sense. This occurs because technologists are occasionally poor strategists who are taken in by the high-tech, fashionable nature of cells and systems and led to believe that FMS, for example, is the cure-all for their company's productivity and profitability problems. Such mistakes can be sobering revelations.

Examining FMS limitations, it can be seen that:

1. FMS is not a cure-all for productivity and profitability problems. It cannot make order out of chaos, but it can make a major contribution after manufacturing chaos is resolved first. Misapplication is a serious threat, and FMS will not solve the problems of a mismanaged or poorly organized plant. New automated techniques, including FMS, should be applied only to a successful nonautomated activity. Information flow must be efficient before computers are introduced. Material flow must be efficient before automated guided vehicles (AGVs) are acquired.

2. Applying FMS may not be as productive as efforts to outsource or subcontract component manufacturing or to maximize the efficiency of existing workers and processes. Increased machine utilization, improved quality and part throughput, and reduced inventory are the goals. FMS may result from efforts to obtain these goals.

3. The high cost of FMS may be prohibitive. Many systems can easily cost in excess of $8 to $10 million. Perhaps the purchase of a machining center equipped with pallet shuttle (Figure 2-8) or with an automatic work changer (Figure 2-9) would be sufficient. Other similar machines and cells could be purchased as competence and confidence grew. CNC machines possess numerous automated features, such as sensing and probing capability, adaptive control, tool monitoring, and advanced diagnostics. These advanced features coupled with an 8- to 16-hour queue of work to operate unattended during the night, for example, may be all that is required to obtain a partial FMS, and for a fraction of the cost.

4. Purchase of an FMS is not the same as that of a standard commodity-type machine tool even if the system is a turnkey purchase. It is an ongoing partnership venture between supplier and user that can take in some cases 2 years from initial facility preparation to complete system implementation and operation. An FMS purchase evolves from conception and birth through adolescence to maturity.

5. User expectations may be distorted based on preconceived notions as to what FMS is and what it can do. User expectations should be clarified through pursuit of education and information about FMS from a variety of sources.

6. Management commitment may not be perceived as real or genuine even though funding is approved. High-level management accessibility, visibility, and involvement must be seen and sensed by all support personnel. Implementation and operation accountability should be controlled through careful personnel selection and project management.

Figure 2-8   Machining center equipped with pallet shuttle for better machine utilization. (Courtesy of Cincinnati Milacron)

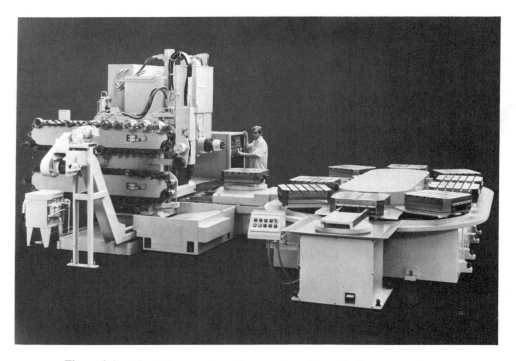

Figure 2-9   Machining center with an automatic work changer can be an alternative to or a beginning for an automated system. (Courtesy of Cincinnati Milacron)

7. Most FMSs are being promoted by machine tool manufacturers who are concerned with metal-removal processes. Closer study of individual component's functional requirements and design criteria may reveal that a material substitution (plastics, ceramics, or the like) or a "chipless" process such as cold forming or precision castings may provide a more practical and economical solution.

8. Purchase of an FMS without consideration as to how the system can integrate (hardware and software) with existing machinery, computer systems, and operations causes automation isolation. Essentially, an expensive "island of auto-mation" can be internally productive but externally unproductive because the system is not or cannot be integrated with other existing operations or systems.

9. System sizing or scoping, relative to workpiece type, configuration, and complexity projected to run through the system, can be either misapplied or inadequate. FMS should not be at the mercy of the workpiece. FMS is for the workpiece—not the workpiece for FMS.

FMS JUSTIFICATION

Intensified global competition is leading many companies toward a renewed com-mitment to excellence in manufacturing. Attention to product and process im-provements gained through high-tech automation has made manufacturing once again a key element in the strategies of companies intending to be world-class competitors. The concept of justification, however, remains a major obstacle to the lasting success of this revolution in manufacturing innovation and capability.

The traditional cost–benefit formulas that may have served business in the past are for the most part out of date. They are too short-term and bottom-line oriented to accommodate long-term and innovative approaches that must be ap-plied to large-scale automation programs like FMS. However, most companies still use the cost accounting and justification practices that were developed decades ago for a competitive business environment drastically different from that of today. In the past, a piece of capital equipment could be justified by the number of people on the line it replaced (direct labor). To install a new machining center that did the work of five people, for instance, meant comparing their salaries plus fringe benefits to the cost of the machine. In this manner, the equipment's purchase was easily justified. Return on investment (ROI) was the driving factor. If a project had a good chance of returning 20 to 30 percent, it had a good chance of being funded.

Traditional justification techniques, based on ROI and direct labor cost re-duction, cannot accommodate the cost avoidance and productivity enhancement achievable with flexible automation systems like FMS. Financial people are using formulas and accounting forms that include only traditional or standard line items and benefits to run the numbers and as a result are delaying innovative and cost-effective automation implementations.

Nontraditional accounting practices must be developed and used that quantify the total impact of automation on the entire business and reflect the broader FMS benefits, for example, reduced inventory, improved quality, reduced scrap and rework, and increased part throughput. Automation investments represent strategic decisions with far-reaching competitive implications and should be evaluated as such. Financial analysts and cost accountants must be able to work with manufacturing engineers to project, incorporate, and justify benefits that accrue business wide in calculations that previously were not considered at all. Their field of vision and measurements must take into account the broad benefits of automation over the long haul.

Several additional barriers exist that hinder the adoption of new manufacturing technologies like FMS and justifying their expenditure. And many of the biggest barriers are managerial, not technical. These include:

1. Excessively High Hurdle Rates

To get a project over the justification hurdle, the initiator inflates the benefits and deflates the costs. The project evaluator (management) has little or no technical expertise with which to evaluate the project and is concerned about being blamed for a failure. The safe course for the evaluator is to raise the hurdle. The winner, of course, is the competition.

2. Outmoded Executive Compensation Plans

These are often the real reasons that underlie the cost-justification roadblock. Companies clearly embarking on a course of technological change and moving to automated cells, FMS, and CIM ought to have their management compensation plans reflect it. If management is not rewarded for change, then how can anyone expect them to cause it to happen?

3. Status Quo Comparison

Most new investments are measured against a status quo alternative of making no investment at all. Such a comparison usually assumes continuation of existing market share, a stable selling price, and consistent related costs. Experience has proved that the status quo does not continue undisturbed and that doing nothing is really falling behind.

4. Incremental versus Turnkey Projects

In many companies the capital equipment authorization process specifies different levels of approval for different amounts of funding. Small investments (under $100,000 for example) may need only the approval of the plant manager. A several million dollar FMS may require the board of directors' approval. Consequently, this apparently reasonable procedure creates an incentive for managers to approve small projects that fall just below the cutoff point. This in turn delivers incremental savings that, over time, can add up to be technologically less but financially more than a complete system. And the time to implement such incremental automation investments can obsolete the earlier equipment, processes, or material.

But perhaps the decades old accounting practices are just being misapplied. Or the root cause of failure to update the financial measurement systems to align with the new technologies is more a matter of management attitude, approach, and business sense than mere number manipulation. Figure 2-10 summarizes the ten sure steps to justification failure for any automation program.

Sound FMS justification should begin with a defined plan or road map as to how to reach the intended automation goal. This plan, in the form of a flow chart in Figure 2-11, should be agreed on and followed by the manufacturing engineers and the financial evaluators working together to try to justify the cell or system. Guidelines and ground rules should then be established for financial measurement and justification, and factors should be developed to effectively measure the new system's performance criteria. Each factor then should be assigned a specific weight or point value depending on its overall importance to system performance and overall effect on product cost, quality, and delivery. The weighted value for each factor to be considered can have a specific dollar value associated with it, which later can be totaled and compared as savings against the systems' actual cost.

Generally, the principal factors related to a winning FMS justification plan include:

1. Incorporate Strategic Long-term Business Objectives into the Capital Budgeting Process

Identifying principal long-term business objectives along with cost-reduction improvements is the first priority. FMS costs should be justified as a strategic investment in automation, similar to an investment made in a child's early education. The three general types of improvements that can be made are:

- ☐ Cost-reduction improvements
- ☐ Quality improvements
- ☐ Delivery (flexibility) improvements

Improved competitiveness comes from decreasing current costs, increasing quality, and improving delivery (responsiveness) time. Cost categories that can be decreased include:

- ☐ Labor
- ☐ Capital
- ☐ Materials

Reducing costs as part of a company's long-term strategic goals and business objectives means more than seeing costs as just expenses. There are costs associ-

---

TEN SURE STEPS TO JUSTIFICATION FAILURE

1. Focus on justifying separate incremental projects rather than on strategies to fit the long-term business plan.

2. Keep management measurement and reward systems geared to obtaining short-term results, with no incentive for initiating change and taking a new approach.

3. Maintain the status quo. It will keep us at least as good as we are today.

4. Make no attempt to establish a justification plan and ground rules between manufacturing engineers and accountants prior to beginning the justification process.

5. Maintain overly conservative capital budget limitations and underestimate resources available for productive investment.

6. Continue to use and not update unrealistic hurdle, discount, and payback rates.

7. Generate and believe overly optimistic sales forecasts reflecting unrealistic market assumptions, demands, and the competition.

8. Continue to overlook indirect and intangible cost savings because if it can't be quantified, it can't be considered.

9. Always let quantitative analysis and payback overrule business judgment and risk.

10. Have inadequate provisions to monitor and audit the automation project.

Figure 2-10   Outmoded accounting and management practices continue to plague the financial justification process.

ated with lost revenue, for example. By making the proper investment in FMS, factors such as quality and customer service increase, and these should be incorporated into any automation justification plan. Once the long-term business objectives are developed and agreed on, justification to support the strategy should be emphasized, rather than justification of each project.

2. Look Hard at Inventory Costs

The greatest cost reductions come from reducing inventory at all levels. Inventory is one of the largest items on any balance sheet. Therefore, to the extent that inventory can be reduced, cash is made available for other purposes. Because FMS and other automation projects reduce plant-wide manufacturing cycles, these reductions free up large amounts of cash that were previously tied up in inventory.

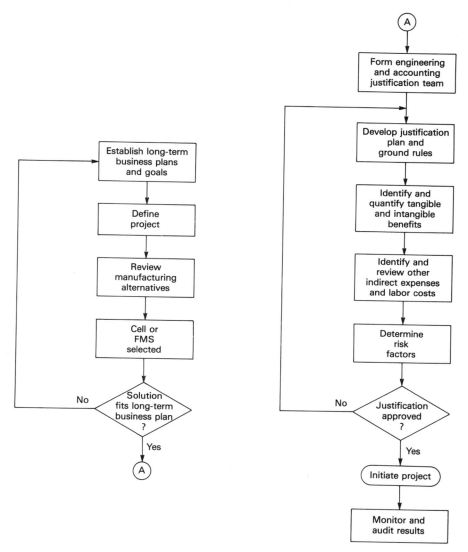

Figure 2-11   Basic cell/FMS justification flow chart.

This in effect becomes the down payment for the automation project. And, by improving material throughput, the time period the material is owned is reduced and therefore the investment in inventory.

3.   Determine Risk Factors

Risk factors weight cash flow depending on the likelihood that costs and benefits will occur in the amount and at the time that they are estimated to. Projects like FMS, depending on system size and complexity, should be grouped prior to assigning risk into broad risk categories, such as:

- ☐ Low risk: high probability of technical success
- ☐ Moderate risk: some risk involved, average chance of technical success
- ☐ High risk: low probability of technical success

Determining risk factors helps to establish a project's relative risk to the financial evaluators once the categories are understood and the general criteria established for the distinction.

4. Measure Intangible Benefits

FMS intangible benefits can directly influence the justification evaluation process if qualitative performance criteria are established before hand. The criteria should be related to the system's strategic requirements and the company's long-term business objectives. Items include improved reliability, responsiveness, and product and process quality, and reduced turnover (better quality of work life). FMS produces bonus ripple effects, such as higher worker productivity from increased job satisfaction and higher market share that comes from reduced lead times. When all the phases are complete, the total system benefits are larger than the sum of its individual parts.

5. Review Other Indirect Expenses and Labor Costs

Close review of indirect manufacturing expenses and associated labor costs is also an important area to consider for justification and savings. This includes, for example, elimination of fork trucks, fork truck drivers, material handlers, laborers, and expeditors. And FMS frequently reduces the number of machines required in a factory, while increasing considerably asset utilization of system equipment. A direct result of this is a reduction in maintenance, time, and costs. These are real measurable savings. Fewer machines also mean less floor space required. This in turn drives down such related expenses as heat, light, taxes, and insurance. Additionally, automation has a strong positive impact on quality, which directly affects scrap, rework, warranty claims, quality-control inspectors, and the postsale service staff. Too often the only labor costs considered for reduction are those that occur on the shop floor. However, other labor costs to be reviewed include engineering, accounting, supervision, and personnel. These are usually treated as overhead costs.

Obviously, more elaborate financial work is involved (depreciation, amortization, discounted cash flow to name a few) to justify FMS or any other automation project than the brief overview presented here. However, the attempt here is to identify some new factors, methods, and practices that can be used to help justify high-cost, high-performance systems.

FMS and every other major manufacturing investment decision involve acknowledging that manufacturing's goal is not only to minimize costs, but also to maximize profits. And increased profits come not only from cost reductions, but also from quality and delivery improvements and enhanced market position. Financial justification should not restrict the adoption of industrial automation so vital to manufacturing competitiveness. By tying justification directly back to the manufacturing operation and the intangible benefits through measurable systems developed and used by manufacturing engineers and financial evaluators, financial risk can be accurately defined and minimized.

MANAGEMENT COMMITMENT AND PLANNING

Any discussion involving small- or large-scale FMS systems or full factory automation programs invokes questions and concerns relative to management commitment and planning. To a large degree, more is at stake than mere financial approval. The terms "commitment" and "planning," although somewhat nebulous, involve depth, breadth, and employee perception issues.

Although more detailed FMS planning elements will be discussed later, it is important at this stage to understand how management commitment and planning affect the overall success of any automation improvement program. Commitment and planning go hand in hand; you cannot be committed to automation without doing the required planning, and you cannot accomplish the intracacies of detailed planning without commitment.

The adopting of FMS and other automation technologies in the United States has been relatively slow, much as the adoption of numerical control has been slow to progress. Disregarding the financial and justification issues of adoption, part of the problem lies in the fact that automation decisions, whether mandated by a top-down approach or evolved through a middle management, grassroots, bottom-up effort, are made without full regard to how those decisions must be backed, led, and precisely understood. Automation improvement decisions like FMS are generally made at executive levels, with implementation responsibility and accountability being delegated to lower organizational levels without full regard to how senior management can help, other than by monitoring the entire process.

Much of U.S. industry's executive management come from financial, legal, and marketing types of backgrounds, and many are for the most part "technologically aversive." Hence, some lack the necessary production, manufacturing, and general operating experience to ask the right probing questions, while at the same time helping, advising, and leading the entire modernization effort.

Highly successful automation projects are more likely to occur in companies that plan from the top down and implement from the bottom up. "Islands of automation" and "Band-Aid" automation fixes can be dangerous and risky and may well turn a company against automated technology in general.

Planning is a distributed decision-making process. It involves top management for leadership, direction, judgment, major decision making, and removing roadblocks; middle management for implementing change, carrying out decisions, and managing results; and producers for doing the work and providing information, insight, and knowledge.

Planning works best when it deals with current issues and the foreseeable future and when it outlines immediate actions and benchmarked results. Detailed elements of planning involve some aspects that are more easily planned and the results benchmarked and measured than for others. These elements range over planning changes covering both visible and invisible aspects of planning. Visible elements, such as planning facility changes and modifications to accommodate FMS, for example, tend to be easier to scope out than how the FMS will interface with existing computer systems and current operations.

Once a major automation project is started, for example, it can easily take a few years to complete. Allowing for anticipated changes in priorities and people during that time is only taking a realistic view of a changing world and another element of the planning process. Consequently, even change itself must be planned for and dealt with. Planning for change can be accomplished by open discourse, astute contingency planning, and designating human resource backups and by management becoming more involved in the planning process. Management involvement provides the necessary "glue" through communication, support, direction, and leadership to keep the project tied together, moving, and on schedule.

What can management do to exhibit more involvement in the cell or system installation? How can they better support and plan for their company's success through factory automation efforts? Elements of commitment and planning take their form from many different perspectives. Generally, management's responsibilities, along with visible signs of the commitment and planning effort, should be:

1. Management must initially concentrate on the "where we are going" issues and be accessible and approachable by the project team. Management must be available to provide guidance and direction, while helping to seek answers to questions that they themselves cannot answer. Simple, once a month summary review meetings are not enough. Management must keep its finger on the pulse of the emerging automation improvements, just as on the cost, quality, and delivery of products. Senior management visibility and accessibility by middle, first-line management and hourly employees are vital to overall attitude changes and project success before, during, and after automation improvements.

2. Plant charters and mission statements are important but are quickly perceived as mere lip service unless backed by concrete, tangible evidence of FMS improvement efforts. People must see short-term deliverables take shape toward completion of the FMS. Or, as the saying goes, "seeing is believing."

3. Commitment means communication and reassurance, not only to members of the project team, but to all employees, particularly those who feel most threatened by automation in the first place. Management should get out of their offices, walk around, and talk to people. This shows sincerity, individual concern, and adds the personal touch. And senior managers may be surprised at what they learn.

4. Management must surround themselves with strong, competent people. Going one step farther, this means adding balance to the project team in terms of skills, disciplines, knowledge, and experiences. Cronyism, political appointments, and recruiting all "yes" people have no place in an FMS or any automation environment. In some cases it may pay to balance the project team with some younger, less experienced engineers and technicians. What many young people lack in broad knowledge and experience, they more than make up for in enthusiasm, drive, and determination, while maintaining a questioning attitude. People must challenge and be challenged for the project to win.

5. Many offshore competitors are successful because they have the ability to function as a group. Management must focus effort on project teams and groups functioning as a whole and eliminate departmental and territorial protectionist thinking.

6. Management must accept the fact that in order to implement FMS many operational and organizational changes may be required, including changes possibly affecting their own jobs. Management must take the responsibility to orchestrate the necessary changes organizationally, operationally, and procedurally.

7. To counteract resistance to change, management must make sure all employees readily see senior-level support and dedication to FMS through regular meetings. These meetings should be open, informal, question and answer get-togethers that stress the benefits, advantages, and anticipated results of the automation effort. Additionally, they should stress the importance of each employee's contribution toward that effort.

8. Bringing in outside consultants to assist or advise with FMS or other automation improvement plans and implementation may be management's way of ensuring success that "we're on the right track." Translated into middle-management, engineering and technician terms, this thinking is very often interpreted as:

☐ Top management doesn't trust our judgment, capabilities, and recommendations.

☐ They don't believe us.

External consultants, in many cases, are very competent and capable individuals with good advice and assistance to offer. However, they should be avoided if at all possible, as their use just adds to the communication, credibility, and lack of trust issues. A strong internal FMS user team should be formed to become educated and work closely with a strong and reputable supplier team in a partnership approach to FMS.

SUMMARY

1. Conventional approaches to manufacturing have generally centered around machines layed out in arrangements by function, line, or cell.

2. Problems associated with conventional manufacturing include both external and internal pressures and inefficiencies.

3. The principal benefits derived from implementing FMS are inventory reduction, direct labor savings, increased asset utilization, and reduced floor space requirements.

4. Most firms that have installed FMS have done so on the basis of reduced inventory, increased part throughput, predictable quality, and improved responsiveness.

5. Many FMS limitations originate from unrealistic expectations as to what FMS is and what it can do and misunderstanding about the need for flexibility in manufacturing.

6. The need for FMS is not as important as the need for flexibility in manufacturing.

7. Traditional justification techniques, based on ROI and direct labor cost reduction, cannot accommodate the cost avoidance and productivity enhancements achievable with flexible automation systems like FMS.

8. Nontraditional accounting practices that quantify the total impact of automation on the entire business and reflect the broader FMS benefits must be developed and used to justify FMS.

9. Manufacturing engineers and accounting personnel must be able to work together as a justification team to plan, project, incorporate, and measure benefits that accrue business wide in calculations that previously were not considered at all.

10. The principal factors related to a winning FMS justification plan are to incorporate strategic long-term business objectives into the capital budgeting process, to look hard at inventory costs, to determine risk factors, to measure intangible benefits, and to review other indirect expenses and labor costs.

11. Highly successful automation projects are more likely to occur in companies that plan from the top down and implement from the bottom up.

12. Planning works best when it deals with current issues and the foreseeable future and when it outlines immediate actions and benchmarked results.

13. Management's commitment and involvement provide the necessary glue through communication, support, direction, and leadership to keep the project tied together, moving, and on schedule.

REVIEW QUESTIONS

1. A machine layout by _____ is the arrangement of similar machines by department.

2. A _____ layout is an interdepartment trade-off for intradepartment material handling.

3. Which of the following is an external conventional manufacturing pressure or inefficiency?
   (a) Extremely long lead times
   (b) Inadequate or unavailable tooling
   (c) Uncertainty and instability of economic conditions
   (d) Excessive expediting as a result of front-loaded lateness

4. Direct labor savings are the most important benefit derived from implementing FMS. True or False?

5. Material costs account for only about 10 to 15 percent of most products. True or False?

6. _____ can be broadly defined as the ability to respond effectively to changing circumstances.

7. Name and describe FMS limitation factors.

8. Which of the following is not considered a major barrier to justifying new manufacturing technologies?
   (a) Excessively high hurdle rates
   (b) Traditional and outdated cost–benefit formulas
   (c) The status quo comparison
   (d) Lack of technical and financial know-how

9. Letting quantitative analysis and payback always overrule business judgment and risk is the best rule to follow for a sound justification plan. True or False?

10. The three general types of manufacturing improvements that can be made are cost reduction, delivery, and _____ .

11. Cost-reduction categories that can be decreased in a manufacturing operation are labor, capital, and _____ .

12. Incorporating strategic long-term business objectives into the capital budgeting process is the first step toward establishing a winning FMS justification plan. True or False?

13. _____ expenses such as elimination of fork trucks, fork truck drivers, laborers, and expeditors should be closely reviewed as part of an FMS justification plan.

14. Highly successful automation projects are more likely to occur in companies that plan from the top down and implement from the bottom up. True or False?

15. Discuss the steps management can take to show visible signs of commitment to a major automation program like FMS.

3

Manufacturing Cells

OBJECTIVES

After studying this unit, you will be able to:

☐ Classify manufacturing cells into four general categories

☐ Describe the concept of unattended machining

☐ Identify some unattended turning center and machining center features and requirements

☐ List and discuss the primary differences between an automated manufacturing cell and a true FMS

INTRODUCTION

The concept of processing parts in a manufacturing cell is usually developed around NC machines as the core equipment. Machining and manufacturing cells can range in size and complexity from an automated stand-alone turning or machining center to a full-scale FMS. The terms "cell" and "FMS" are sometimes used interchangeably; however, many differences exist depending on system size, complexity, and the extent of the application, along with how a particular company views and uses its system.

Machining cells are generally based on clustering of two or more redundant CNC machines by part family. Part loading and unloading within the cell are usually handled by means of an integrated, computer-controlled robot or some other partially automated or manual means.

Many companies are overcoming their apprehension of cells and unattended machining problems by seeing the extra production they can get from grouping their machines and setting them up to run during lunch hours and for a few hours after quitting time. Many industrial managers have come to recognize and accept the need for greater automation coupled with greater flexibility in their manufacturing operation through the use of unattended or lightly attended cells—and for a fraction of the cost of a full-scale FMS.

CELL DESCRIPTION AND CLASSIFICATIONS

As discussed in Chapter 2, a definition of manufacturing cells in its broadest sense implies the logical arrangement of stand-alone manual or NC equipment into groups or clusters of machines to process parts by part family. Additionally, by definition, processing parts in a manufacturing cell includes completing as much of the workpiece processing as possible within the cell before moving it to the next sequential processing, stocking, inspection, or assembly station. Today, the term "manufacturing cell" is much broader and further implies some level of automated part loading, unloading, delivery, or exchange to the clustered machines. And yet the term manufacturing cell can take on a variety of different meanings depending on a manufacturer's application, state of technological development and understanding, and particular point of view.

Basically, manufacturing cells can be divided into four general categories:

1. Traditional stand-alone NC machine tool
2. Single NC machine cell or minicell
3. Integrated multimachine cell
4. FMS

The stand-alone NC machine tool (Figure 3-1) is characterized as a limited-storage, automatic tool changer and is traditionally operated on a one-to-one machine to operator ratio. It has been the mainstay of capital acquisitions in the metal-cutting industry since the mid 1960s and continues to generate the bulk of CNC machine tool sales.

In many cases, stand-alone NC machine tools have been grouped together in a conventional part family manufacturing cell arrangement (Figure 3-2) but still operating on a one-to-one machine to operator ratio. Machines within a cell of this type have sometimes been painted a similar color to further add cell distinction for a particular group of parts and differentiate it from other cells (for example, red cell, green cell, blue cell).

Some stand-alone NC machines are characterized and operated as a cell by virtue of a change in the machine to operator ratio. These machines are usually redundant and are operated on a two-to-one (Figure 3-3) or in some cases a

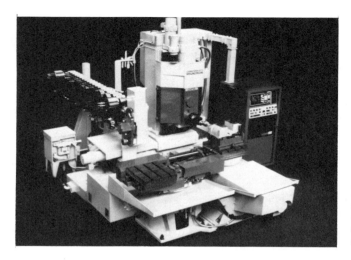

Figure 3-1 The traditional stand-alone NC machine tool is typically operated on a 1 to 1 machine to operator ratio. (Courtesy of Cincinnati Milacron)

Figure 3-2 Some stand-alone NC machine tools have been grouped together in a conventional part family cell arrangement, but still operating on a 1 to 1 machine to operator ratio. (Courtesy of Cincinnati Milacron)

Figure 3-3   Some NC machines are grouped as a cell and operated on a 2 to 1 or 3 to 1 machine to operator ratio. (Courtesy of Cincinnati Milacron)

three-to-one machine to operator ratio. Because one operator is running more than one machine, the group of machines one operator is therefore responsible for is sometimes referred to as a cell. Part cycle start and stop times in a cell of this type are controlled and staggered by the operator so that one machine is always running while the other is idled for loading or unloading.

The single NC machine cell (Figures 3-4 and 3-5) is characterized by an automatic work changer with permanently assigned work pallets or a conveyor–robot arm system mounted to the front of the machine, plus the availability of bulk tool storage. There are many machines with a variety of options, such as automatic probing, broken tool detection, and high-pressure coolant control, that fall into this category. The addition of these and other special option features enables many single NC machines to operate as a self-contained cell. The single NC machine cell or minicell is rapidly gaining in popularity, functionality, and affordability because it can be purchased for a fraction of the cost of a complete FMS and can be programmed and loaded with parts to run unattended for several hours. Unattended operation of a single machine cell affords users the opportunity to gain some of the advantages of an FMS, such as increased spindle utilization and reduced direct labor, while increasing their knowledge and confidence about unattended machining and automated manufacturing capabilities.

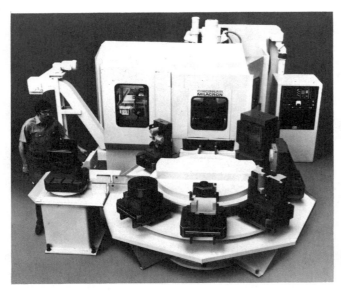

**Figure 3-4   Single NC machine cell
with a permanently assigned and
attached pallet work-holding
system. (Courtesy of Cincinnati
Milacron)**

The integrated multimachine cell is made up of a multiplicity of metal-cutting machine tools, typically all of the same type, which have a queue of parts, either at the entry of the cell or in front of each machine. Multimachine cells are either serviced by a material-handling robot (Figure 3-6) or parts are palletized in a two- or three-machine, in-line system (Figure 3-7) for progressive movement from one

**Figure 3-5   Another single NC machine cell with a conveyor and robot
arm to automatically deliver, load, and unload turning center parts. (Courtesy of Cincinnati Milacron)**

Figure 3-6 A multimachine cell may be serviced by a material-handling robot. (Courtesy of Cincinnati Milacron)

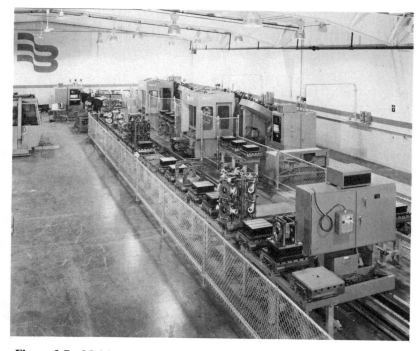

Figure 3-7 Multimachine cells can be composed of different machines and linked together by an in-line material-handling system for progressive machining at each station. (Courtesy of Kearney and Trecker Corp.)

machining station to another. The typical application of a multimachine cell serviced by a robot (Figure 3-8) is high-volume production of a small, well-defined, design-stable family of parts. Machines can be different in a cell of this type, and workpieces can be progressively moved, for example, from a turning center to a center-type grinder for part completion. Palletized in-line cells can also be applied to either high- or low-variety and -volume production applications. Material handling links together a group of flexible general-purpose machine tools utilizing a common pallet design with prefixtured parts on pallets.

The FMS, sometimes referred to as a flexible manufacturing cell (FMC), is characterized by multiple machines, automated random movement of palletized parts to and from processing stations, and central computer control with sophisticated command-driven software. The distinguishing characteristics of this cell (Figure 3-9) are the automated flow of raw material to the cell, complete machining of the part, part washing, drying, and inspection within the cell, and removal of the finished part.

New technology, if implemented too quickly and for technology's sake, can

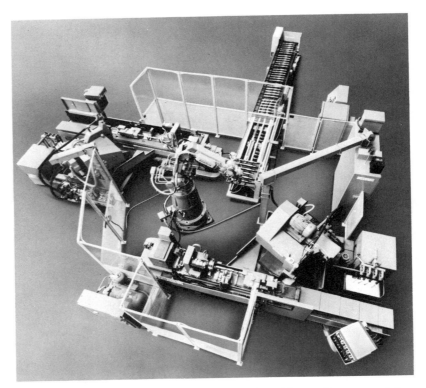

Figure 3-8 Typical application of a multimachine cell: a material-handling robot progressively moving a part from machine to machine. (Courtesy of Cincinnati Milacron)

seriously affect a company's productivity and shock an organization with its functional demands, requirements, and unforgiving consequences. As a result, many companies, feeling that a single or integrated multimachine cell is too little but a full-scale FMS is too much (in terms of cost, technology, and risk), elect to take a phased approach to system implementation. The broader the scope and span of a company's manufacturing objectives and workpiece requirements are the greater the need for a phased system installation and implementation.

Developing a master FMS or system plan and then installing cellular pieces of the system to be linked together at a later time is a logical and chosen approach for many companies. FMS and cellular systems can replace existing production techniques for an entire part family and may be implemented over a period of time. A phased approach allows the rest of the organization and human resources to adjust and adapt to the new requirements and demands of the system. This is best accomplished by developing a comprehensive plan for phases of productivity improvements that match market and product requirements. A phased FMS installation based on a beginning cellular approach to manufacturing gives people learning time to utilize and maximize the system's productive capabilities.

UNATTENDED MACHINING

The concept of unattended machining implies running an NC machine tool with no operator in attendance for extended periods of time, usually eight or more hours. The parts, tools, and NC programs are considered to be loaded and available at each machine station or are delivered on an as needed basis to each machine.

Figure 3-9   Flexible automated material flow is a distinctive feature of FMS. (Courtesy of Remington Arms)

For many manufacturers, the purchase of a new single or multimachine cell is their first attempt at unattended machining. This, however, does not have to be the case as considerable learning can take place in attempting to operate any existing NC machine tool unattended. Then, by the time a newly purchased single or multiple machine cell is installed, the responsible people will have accumulated extensive knowledge and information about the many start-up problems associated with unattended machining.

Unattended machining begins by making sure an adequate supply of parts and cutting tools is available to keep the machine in operation for an extended period of time. Questions need to be asked and answered relative to the unattended operation, such as:

1. What type of material are the parts to be produced from? Steel? Cast iron? Aluminum? A combination? Steel parts are usually harder to work with than cast-iron parts because of the stringy, curling chips produced by machining steel.
2. How long will the machine continue to operate unattended?
3. What provisions are available to keep it running?
4. What assurance is there that the machine will continue to run and not cause damage to itself, the parts, or the tooling?
5. How many different types of parts can be run unattended on the machine or in the cell?
6. How much extra process preparation work is required ahead of time to process each type of unattended part (part programs, fixtures, tooling, or material-handling changes)?

The development of unattended machining is really a matter of expanding the scope of machine applicability, improving the assurance that the run of parts will be completed, avoiding wrecks, and making the entire operation user friendly.

Newly purchased single or multimachine cells should include features such as extended part queuing and tool-changing capabilities (Figure 3-10), torque and force sensing of the metal-cutting process (Figure 3-11), automatic fault detection, and probing capabilities (Figure 3-12). These expand and make possible the overall scope of unattended machining, without which the length of any unattended machining continuous run would be considerably limited.

The benefits of unattended machining are essentially the same as a cell or a subset of an FMS. These include increased machine utilization, improved quality through increased consistency and predictability of operation, reduced floor space because fewer machines are required to make the same production, and reduced direct labor. Work-in-process inventory, however, will generally not be reduced with the installation of one unattended machining cell. In fact, a single NC machine cell put in the middle of a batch manufacturing operation may actually increase work in process depending on how parts are arranged and scheduled for unat-

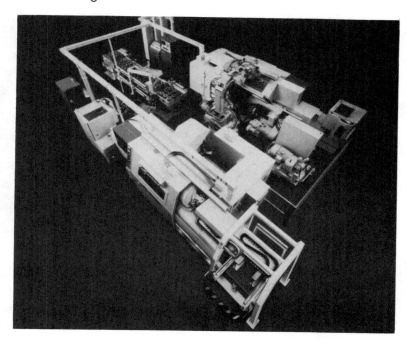

Figure 3-10 Single or multimachine cells include extended tool-changing capabilities to maximize unattended operation. (Courtesy of Cincinnati Milacron)

Figure 3-11 Torque and force sensing of the metal-cutting process is a vital feature for unattended machining. (Courtesy of Cincinnati Milacron)

Figure 3-12 **Part probing greatly contributes to accuracy, repeatability, and quality of unattended machining. (Courtesy of Cincinnati Milacron)**

tended machining. Unless the entire manufacturing flow and process is altered to accommodate an unattended cellular machining application, any real work-in-process inventory reductions will not be realized.

Added costs will occur in the indirect areas of methods engineering and part programming. This is due to the extra thinking and planning time that must be applied in the areas of tooling, fixturing, methods, and process planning. And additional logic must be built into each NC part program for probing cycles, torque and force sensing, broken tool detection, and other programmable features. Any unattended cellular or FMS machining application will require additional effort in these indirect areas in order to take full advantage of available equipment options and features.

Unattended machining is generally attempted with single or multimachine cells consisting of turning centers or machining centers. Although some unattended machining similarities exist between turning centers and machining centers, each has its own application features and requirements.

Unattended turning center features and requirements are:

1. Part size must be controlled through probe measurement of the part, automated in-process or postprocess gaging, and automatic compensation of the machine for changes.

2. Parts must be automatically delivered, loaded, and unloaded, usually by means of an integrated floor- or machine-mounted robot arm along with part queuing by palletized conveyor.

3. Specific part identification can occur by probing unique dimensions to distinguish random parts within a given family and calling up the proper NC program as required.

4. Wrecks can be avoided through spindle torque and slide force sensing and shutting the machine down safely before part, tool, or machine damage occurs. The primary reasons for machine shutdown in an unattended application are mainly due to exceeded machine horsepower limits, dull tools, or excess workpiece stock.

5. Some chips can be cleared from the chuck by the quick rotation of an empty chuck and applying an air blast before loading the next part. However, chip control is a general problem (Figure 3-13) in unattended applications, which includes the breakage of chips to keep them from getting caught in the tool change mechanism, interference with probe cycles, and other post- or in-process inspection gaging equipment. And long unattended applications can create a problem with disposal of the volume of chips produced.

6. Parts must be easily turned end for end and accurately located for part completion. This is an important requirement if the full advantages of an unattended turning cell are to be realized.

7. A fixed probe for automatic tool length setting (Figure 3-14) avoids the time-consuming manual ordeal of tool setting. It can also be used to measure and compensate for minor tool point changes in position during machine operation for such things as threading tools, where the part cannot be measured by a turret probe.

Unattended machining center features and requirements include:

1. Palletized queuing of workpieces by means of a carrousel or in-line part setup arrangement

2. Detection of missing, misaligned, or wrong parts using the probe and part program branching to bypass the problem areas

3. Determination of excess or insufficient stock with the probe

4. Detection of broken tools through the use of a fixed probe or with a spindle probe (Figure 3-15) to determine whether the workpiece stock has been removed

5. Avoiding erroneous probe data, such as touching on a chip with a reasonableness test (should be approximately within these limits) on the probe data

6. Avoid wrecks by spindle torque measurement and predetermined shutdown limits

Figure 3-13   Depending on material type, chip flow, control, and volume can be a problem in unattended machining. (Courtesy of Cincinnati Milacron)

Figure 3-14   Fixed probes can be used for automatic tool length setting and for broken tool detection. (Courtesy of Mazak Corp.)

7. Clear intermittent and "bird's-nest" chips (long stringy steel chips tightly bundled together) from tools before exchanging by pulling away from the workpiece and spinning the spindle in reverse, for example
8. Clear chips for probing cycles through intermittent high-pressure coolant flushing and air blasts

Although manufacturing requirements are rapidly changing and intensifying, unattended machining features continue to be improved and developed for both turning and machining center cellular applications. However, to make use of these developing unattended machining features, more attention, effort, and skill will be required in the process planning, tooling, and part programming phases of the manufacturing cycle.

CELLULAR VERSUS FLEXIBLE MANUFACTURING

An FMS, even though it is a unique manufacturing system, is sometimes referred to as a cell—a highly sophisticated and automated one—among some manufacturers. And, in some cases, a single NC machine cell or an integrated multimachine cell is referred to as a flexible manufacturing system (FMS); in actuality, the only "flexibility" that may exist at all is changing to a different part when similar batch requirements are completed. Although sometimes described as a cell, an FMS is set apart from any other cell by virtue of its central computer control and highly developed software; complete part, tooling, and material-handling flexibility and control; and randomness of production scheduling and machining. Conventional equipment cannot compete with FMS concepts because of the lack of management

Figure 3-15 The spindle probe can be used to determine part presence and stock allowances. (Courtesy of Cincinnati Milacron)

control, inherent inefficiencies, and the ever present setup and retooling requirements.

Both similarities and differences exist between cells and systems. Similarities exist from the viewpoint that the level of automation for either a cell or system can vary depending on how much technology and money will be applied and whether the cell or system will be operated unattended or lightly attended. Both cells and systems possess multiple part processing and part program storage capability. Cells can store and access limited but multiple part programs at each machine control unit, while an FMS possesses central computer storage and retrieval capabilities for numerous part programs. Automatic or semiautomatic part loading can be accommodated in either a cell or FMS. Magazine, hopper, guided vehicle, multistation shuttle, and robot are possibilities that part configuration may help determine. Additional similarities exist relative to the fact that completion and prior prove-out of part programs, tooling, and fixturing must be completed before either cell or system implementation. And the extent of the planning that must be done prior to a cell or FMS installation is directly proportional to the size, type, and complexity of the cell or system to be installed. However, the fundamental require-

Figure 3-16   Cells are generally controlled by cell controllers or by their own independent but interfaced machine controllers. (Courtesy of Kearney and Trecker Corp.)

ment with either a cell or FMS is to target the part family or families to be produced in order to scope out and determine the type, size, and configuration of the cell or system.

Planning for either a cell or system involves close and careful analysis of the parts to be produced, along with a thorough review of part and fixture drawings, present routings, setup and tooling sheets, and part programs. From these and other sources, the following detailed data should be obtained and organized: smallest to largest and lightest to heaviest parts to be produced, extent of work-piece material dissimilarity including shape and hardness, dimensional tolerance and surface finish requirements, analysis of work content and operations required, present or expected lot sizes and frequency of runs, present and expected production rate, range of annual production requirements, and present identifiable problems.

Fundamental questions to be answered include how work is to be delivered to and from the cell or system, whether there are any floor space or elevation restrictions or requirements, and whether there is a central coolant–chip disposal system to tie into or construct.

Planning involves a considerable amount of work whether a cell or system is being considered for purchase. In many cases, once detailed planning has begun the decision criteria relative to a cell, multicell, phased-growth cell, or FMS purchase can change many times over. System, facility, and human resource planning is so important to a cell or system start-up and success that the next two chapters are devoted exclusively to the subject.

The primary differences between an automated manufacturing cell and a true flexible manufacturing system (FMS) are:

1. Cells lack central computer control with real-time routing, load balancing (software), and production scheduling logic. They are generally controlled by cell controllers or by their own independent but interfaced machine controllers (Figure 3-16). An FMS will almost invariably be connected to a higher-level computer (Figure 3-17) within the manufacturing operation. In many cases it is tied directly to the corporate computing system, which may also be running the MRP (material requirements planning) system, the inventory control system, and sometimes the CAD (computer-aided design) system in design engineering.

2. Cells are typically tool capacity constrained. Both single and multimachine cells are limited by the total number of unique and redundant cutting tools that occupy available tool pockets (Figure 3-18). This limits the part spectrum that could be run through a cell at a given time without stopping the equipment and manually exchanging tools to accommodate different workpieces. An FMS with automated tool delivery and tool management can automatically transfer, exchange, and migrate tools through centralized computer control and software independent of equipment activity. With a cellular application, the cutting tool count must be minimized to offset the limited tool buffer storage of the machine. Parts must be closely scrutinized and part prints sometimes changed in order to match the part family tool range with the available tool pockets.

**Figure 3-17   An FMS is controlled
by a higher-level computer due to
the sophisticated software required
to drive the entire system.
(Courtesy of Cincinnati Milacron)**

3. Cells generally have less flexibility than an FMS and are restricted to a
relatively tight family of parts. As long as the part family remains unchanged and
design-stable, the automated cell can operate very efficiently. An FMS, on the
other hand, has greater depth and breadth of flexibility due to the range of parts in

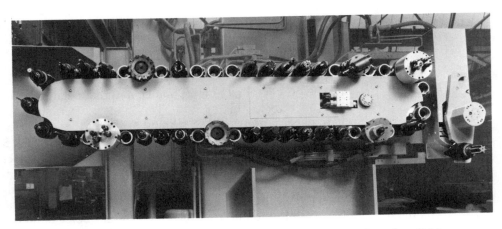

**Figure 3-18   Cells are generally limited by the total number of available
tool pockets on each machine in the cell. (Courtesy of Cincinnati Milacron)**

varying lot sizes that can be accommodated in the system, random machine scheduling, and automated material flow and movement. In some multimachine cells, parts are passed in sequence from one machine to another. Whether material handling is automated or not, this type of cell configuration can present problems when some flexibility of part variation requires certain machining operations to be omitted, added, or changed.

A major <u>disadvantage</u> of isolated cells is a lack of management discipline during times of shop load imbalances. If one cell is underutilized, for example, while others are loaded at or over capacity, a common tendency is to shut one cell down and "stripout" existing machine tools to handle the overcapacity of the other cell or cells. Such crisis management decisions violate the integrity of the cell being stripped and causes problems and confusion again when parts for the stripped cell are required. And people become confused and disillusioned when they have established identity with a cell and see the cell concept abused to fix short-term capacity problems.

Attempting to grow or phase a cell or a series of cells into an FMS can also be a drawback. Unless a well-organized, well-developed master plan is assembled and executed with critical time-phased deliverables, a lengthy piecemeal approach to cell integration will result. This can lead to first-purchased cellular equipment becoming obsolete or lacking integration capabilities by the time the last-purchased equipment is actually bought and installed. This can further lead to part manufacturing conflicts and infighting as requirements change over time and be interpreted by employees as a less-than-serious management attitude toward automation improvements.

Whether a potential user is drawn toward and can afford either a cell or an FMS, it is important to make sure that all the broad differences are clearly understood, as depicted in Figure 3-19, and that the cell or system purchase should be driven by the application-intended and -targeted part spectrum.

CELL	FMS
Low Flexibility	High Flexibility
Small stored part program inventory and accessibility	Large stored part program inventory and accessibility
Limited on-line computing power and decision-making software	High on-line computing power and decision-making software
Low to moderate equipment and resource costs	High equipment and resource costs
Limited flexibility and variety of parts produced	High flexibility and variety of parts produced
Low to medium preparation and implementation requirements	High preparation and implementation requirements
Benefits narrow but easily identified and quantified	Benefits broad but hard to identify and quantify
Moderate justification complexity and difficulty with mid-management approval required	Difficult and complex justification process with high-level approval required
Moderate level of management commitment and support required	High level of management commitment and support required

Figure 3-19   Contrasting principal cell/FMS differences

CELL	FMS
Low staffing and training impact	High staffing and training impact
Moderate effect on other internal operations and organizations	High effect on other internal operations and organizations
Low to moderate risk and complexity, minimal facility changes	High risk and complexity, many facility changes or new facility required
Short planning to implementation cycle	Long planning to implementation cycle
Quick and practical learning curve and implementation cycle	Lengthy and involved learning curve and implementation cycle
Generally no tool delivery; limits system scope and flexibility	Generally tool delivery and tool management; opens systems scope and flexibility
Possibility of violating cell integrity by "stripping out" underutilized equipment to handle excess capacity during times of peak load conditions	Not likely to violate FMS integrity and strip out equipment due to inherent flexibility and centralized computer control of the system
Phased cellular growth may take too long and obsolete equipment before full integration	FMS complete system and total part spectrum planned and implemented in turnkey installation

Figure 3-19   *(Continued)*

SUMMARY

1. Manufacturing cells can be divided into four general categories: (1) stand-alone NC machine, (2) single NC machine cell, (3) integrated multimachine cell, and (4) FMS.

2. The broader the scope and span of company manufacturing objectives and workpiece requirements, the greater the need for a phased system installation and implementation.

3. A phased FMS installation based on a beginning cellular approach to manufacturing gives people learning time to utilize and maximize the system's productive capabilities; however, it can be a drawback if it is spread over too long a period of time.

4. The concept of unattended machining implies making sure that an adequate supply of parts and cutting tools is available to keep the machine in operation with no operator in attendance for an extended period of time.

5. Unattended machining is generally attempted with single or multimachine cells consisting of turning centers and machining centers.

6. The benefits of unattended machining are essentially the same as a cell or a subset of an FMS.

7. With unattended machining, direct labor costs decrease and indirect costs increase as a result of the extra work that must be applied in the areas of tooling, fixturing, process planning, and NC programming.

8. The terms "FMS" and "cell" are sometimes used interchangeably; however, the primary factors that distinguish a cell from an FMS are that cells lack central computer control with sophisticated software, are restricted by a limited number of available tool pockets, and are generally limited to a tightly defined family of parts.

9. A major disadvantage of isolated cells is a lack of management discipline during times of shop load imbalances.

10. The fundamental requirement with either a cell or FMS is to target the part family or families to be produced in order to scope out and determine the size, type, and configuration of the cell or system.

REVIEW QUESTIONS

1. Manufacturing cells are classified into four categories. Name them.
2. Machine tools are always identical in an integrated multimachine cell. True or False?
3. An FMS requires a limited, well-defined, and design-stable family of parts. True or False?

4. Steel parts are usually harder to work with in an unattended machine application than cast iron or aluminum. True or False?

5. The benefits of unattended machining are:
   (a) Difficult for managers to recognize and achieve
   (b) The same as a cell or a subset of an FMS
   (c) Only in the areas of direct labor reduction and improved quality
   (d) Not possible to achieve with a multimachine cell approach

6. A single NC machine cell operating in the middle of a batch manufacturing operation may actually increase work in process inventory. True or False?

7. A practical and automated method of avoiding wrecks during unattended machining is to:
   (a) Slow machining speeds and feeds to 80 percent of optimum
   (b) Keep cutting tools sharp and accurate
   (c) Control stock allowances on raw material
   (d) Sensor monitor spindle torque and slide forces

8. List and discuss three unattended turning center and machining center features and/or requirements.

9. Why is a cell sometimes referred to as an FMS and an FMS sometimes referred to a cell?

10. Cells typically have automated tool delivery. True or False?

11. The primary reasons for machine shutdown in an unattended machining application are:
   (a) Exceeded machine horsepower limits
   (b) Dull tools
   (c) Excess workpiece stock
   (d) (a) and (b)
   (e) (a), (b), and (c)

12. Describe the three primary differences between an automated cell and an FMS.

13. Cells typically have a lower staffing and training impact than flexible manufacturing systems. True or False?

Part II

Systems Planning

Planning for FMS or any factory automation project is a process that demands total commitment, broad-based business sense, keen technical understanding, and strict attention to detail. It involves squarely addressing not only the technical issues, but also human resource effects, cost, quality and delivery concerns, and long-range product and process vision.

The planning process should begin with highly objective preplanning; an inside-out assessment and evaluation of the company's products, past performance, markets, and competition. Without such a feasibility study, it is impossible to identify if FMS is applicable to the business. Such a study will identify company goals and objectives, thereby laying the groundwork for a strong manufacturing plan.

In-depth preplanning must compare the financial and technical feasibility of FMS with the company's short- and long-range objectives and capital expenditure plan. Equally important in the preplanning phase is the objective and realistic assessment of implementation timetables, human resource expertise and attitudes, and in-house technological and engineering capabilities.

The overall planning process must not be underestimated. Sound and detail-oriented planning is the single most important factor in the pursuit of automated manufacturing. Done incorrectly, it can be a costly, unforgiving experience that seriously affects corporate profitability and destroys employee morale and motivation. Done correctly, it can be a profitable, rewarding, and confidence-building undertaking that can leapfrog a company ahead of its competition.

4

Physical Planning

OBJECTIVES

After studying this unit, you will be able to:

- Understand the value of detailed planning to a successful factory automation project.
- Describe principal planning preparation guidelines.
- Identify separate and joint user and supplier responsibilities.
- Discuss the elements of successful user and supplier project teams.
- List the primary factors in determining FMS size, description, and configuration.
- Identify four basic types of an FMS layout.

INTRODUCTION

The physical planning process involves the detailed technical ''how to get there'' issues. These issues involve pulling together a wide variety of individual skills, disciplines, and backgrounds that will only yield positive results through a structured team approach. A structured team approach will provide an environment that fosters the necessary people involvement and accelerate the decision-making process.

Physical planning for FMS or any automation project can involve hundreds of people. All through the planning, implementation, and installation phases, the problem becomes one of structured focus on evaluation, communication, coordination, and resolution. An automation project such as FMS cannot be developed, controlled, and implemented by informal, haphazard committees assembled irregularly on a "catch as catch can" basis. Strong individuals must be appointed to the project team, from competent project manager to capable, authoritative, and decision-making functional managers. These individuals must be prepared to stay on the project until completion.

Fundamental FMS physical planning involves close teamwork between system user and supplier, along with researching, analyzing, information gathering, compiling, and coordinating a vast amount of complex technical issues and data. This includes targeting the specific parts or part families to be produced and extends through facility modification, system layout, configuration, and sizing, to operating and organization considerations, and compatibility and integratability with existing operations and computer systems.

It has been estimated that as much as 25 percent of the gain expected from automation projects like FMS can be achieved by adopting a proper plan. Physical planning forms the foundation on which the success or failure of an FMS installation is built. It must be given careful, close, and dedicated effort.

PLANNING PREPARATION GUIDELINES

The planning process for FMS or any factory automation project is simpler once responsibilities, accountabilities, and the process itself are clearly understood. In general, planning works best when:

1. The individuals doing the work are accountable.
2. Goals are realistic and attainable.
3. The plans themselves are flexible and have built-in contingencies.
4. The principal contributors collectively agree.

A broad-based analysis should be conducted to determine if machining centers with FMS features (cell technology) or FMS is the best answer for automated manufacturing improvements. The chart depicted in Figure 4-1 can help determine present and future automation needs, goals, and objectives.

Once broad-based goals, objectives, and a technical feasibility study have been completed to determine if FMS is the acceptable answer, the next step is to establish some planning preparation guidelines. In general these would be:

1. Form the Project Team

Team members should be carefully selected "functional" managers. They must have responsibility, accountability, and decision-making authority for their particular department or unit. Included in the process is the selection of a strong

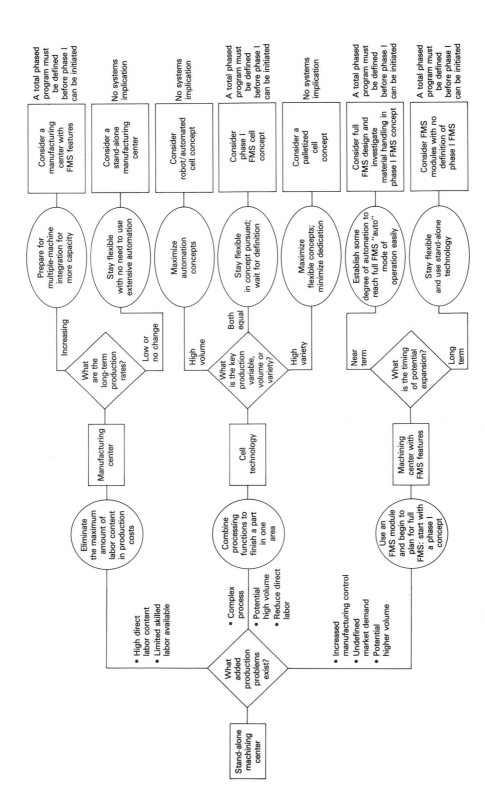

Figure 4-1 A well thought out preliminary plan is important to assess future automation technology needs and to determine if FMS is the proper solution. (Courtesy of Kearney and Trecker Corp.)

project manager. This individual must be part businessperson, part engineer, part lawyer, and part diplomat. This is an individual who is not easily found or developed and careful selection is crucial to project success. The importance of people selection must not be underestimated. The right people must be chosen on the basis of objectivity, knowledge, experience, and proven past performance. They must not be chosen on a "who's available today" basis. Communication lines should be opened early by the project team and should begin with frequent meetings and briefings to inform, educate, and involve. These meetings and briefings should include all the support areas that will ultimately be affected by the FMS installation. This way nobody is left out, and overall project importance is strengthened through top management's accessibility, visibility, and early involvement. Automation is too important and far reaching to be left only to engineers. Decisions and recommendations must be well thought out and multidisciplinary or they will not be accepted.

2. Understand the Process

Team members should understand what is involved in the planning process and the step by step process of FMS planning. This is a fundamental process but one all team members should understand so as not to overlook vital planning elements.

3. Review Goals, Objectives, and Purpose

A functional description of the desired results should be discussed in terms of what the FMS purchase means to company productivity and profitability. Reviewing goals, objectives, and the purpose of a major FMS purchase should be communicated in terms of how it will help the company better address the issues of cost, quality, and delivery, and not in terms of hardware and computer purchases. Goals and objectives should be spelled out in "plain English" by the project manager with top management's participation in this effort. Senior management's involvement at this stage is extremely important to how people see the importance and projected success of the project.

4. Talk to the Vendors

Get potential suppliers involved as early as possible. Involving suppliers early, even at the conceptual development stage, provides expert advice and educational help to the project team, builds team confidence and cohesiveness, and increases chances of success. Early supplier involvement helps to some degree to ensure against finding a better, cheaper, faster, or more efficient solution later on by testing ideas on suppliers. Most vendors are willing to share conceptual ideas at the initial proposal stage. And involving suppliers early in the conceptual stage helps build user–supplier rapport, a necessary ingredient for the teamwork effort required later between user and supplier.

5. Avoid Early Cost–Savings Estimates

Early involvement is necessary for preliminary planning and general concept preparation. Cost and savings estimates should be avoided at this time so as not to establish premature capital investment goals that take priority over productivity improvement gains. Drive toward achieving stated productivity, inventory reduction, machine utilization, and part throughput goals as the first priority. Cost and savings estimates come easier with maximized effort in these areas.

6. Determine User–Supplier Relationship and Approach

The greatest benefits are achieved when a teamwork approach is used. Turning everything over to the supplier in a turnkey approach is a myth and can seriously affect FMS success. The user–supplier team is especially attractive when the supplier can provide products that represent over 50 percent of the system cost. Acting as the prime contractor allows the supplier to control the subsystem vendors, which saves both time and money. The customer then becomes the evaluator, decision maker, and coordinator. The supplier becomes the designer, developer, and prime integrator. Success of the project becomes as important to the supplier as it does to the user.

7. Prepare a Detailed Master Plan

A detailed master plan should be started at this time, ideally set up on a personal computer with one of the many commercially available software scheduling packages. Ideally, an FMS milestone chart (Figure 4-2) should be established. From there, setting up the detailed master plan on a personal computer will permit ease of master plan change and growth as the project matures. It is necessary to establish this master plan very early along the lines of a critical path network, identifying all the many tasks to be performed and by whom, along with start and completion dates. Initially, this master plan is in a very skeletal form. Subcategories and other specific identifiable elements can be added as the project progresses. The master plan should be kept up to date and circulated to all responsible individuals by the project manager. It is *the* controlling document for the entire project; it establishes commitments, responsibilities, deadlines, and activities. It must be respected and rigorously followed.

To obtain the best answers, senior management must participate in the planning preparation process and provide guidance. They must be careful, however, not to give instructions based solely on the business improvement potential without understanding the global engineering complexities and potential pitfalls involved in harnessing the new technology.

Decisions about FMS need to be shared and made by a strong project team able to bring a balanced view to the task at hand. Senior management, operating management, and engineering resources need to be pooled to squarely address questions like the following:

1. Which specific manufacturing areas are most in need of improvement?
2. Which specific improvements will help most in achieving improved productivity, profitability, and product quality, while decreasing costs and lead times?
3. Which of the current manufacturing activities are best suited to current FMS implementation?

FMS MILESTONE EVENTS (PRELIMINARY)		
Item	**Activity**	**Date**
1	Approval of System Layout	January 31, 1990
2	Develop System Requirements	May 23, 1990
3	Site preparation package	June 27, 1990
4	Bid/award site foundation contract	August 22, 1990
5	Five-axis postprocessor completion	October 31, 1990
6	Foundations, utilities, floor slabs complete	January 23, 1991
7	Ship first machining center	October 10, 1990
8	Ship second machining center	October 31, 1990
9	Ship third machining center	December 12, 1990
10	Ship fourth machining center	December 26, 1990
11	Ship fifth machining center	March 20, 1991
12	First machining center installed on temporary foundation	November 21, 1990
13	Second machining center installed on temporary foundation	December 12, 1990
14	Third machining center installed on permanent foundation	February 27, 1991
15	Fourth machining center installed on permanent foundation	April 10, 1991
16	Fifth machining center installed on permanent foundation	May 1, 1991
17	First machining center moved to permanent foundation	May 8, 1991
18	Second machining center moved to permanent foundation	May 15, 1991
19	AGV runoff, local control system integration	August 21, 1991
20	FMS computer system software integration and system runoff	December 4, 1991
21	System acceptance	December 11, 1991

Figure 4-2   The FMS milestone chart provides a master plan for the entire system from planning to operation.

4. What specific but realistically attainable improvements should be targeted for and reached by purchasing an FMS?

5. How aggressive should the implementation plan be? What factors should govern the implementation plan?

6. What manufacturing systems and improvements will the company need to be competitive for the next 10 years?

7. What flexibility should the FMS be able to handle in the way of part volumes, mix, and lot sizes?

8. What are the real manufacturing targets the company should be aiming for?

9. What level of technology is available and viable to achieve the manufacturing targets?

10. Over what period of time should planning through implementation be phased?

These and other difficult to answer questions need to be asked early in the planning phase to help reduce the level of technology risk to which the company may be exposed. Asking questions at the right time is as important as asking the right questions.

THE PROJECT TEAM

Many successful FMS and other large-scale automation projects have been successful because of close attention paid to project team functional relationships, responsibilities, and formation (who does what). Agreement on functional relationships, responsibilities, and assignments in and between the user and supplier organizations early in the preplanning phase can avoid headaches, delays, confusion, and disagreements later.

Project teams in large-scale automation projects such as FMS are successful because of strong program management. Program management is the process of coordinating the achievement of program objectives through the traditional user–supplier organizations and, if necessary, over their unique organizational interests. Program management is used to solve specific identifiable problems with stated performance, cost, and schedule objectives. Additionally, it requires the contributions of many skilled specialists in order to solve problems and remain on schedule throughout the project.

An effectively established matrix management program establishes commitments for program tasks and identifies who is responsible for what and when. Program organizations are utilized only for the completion of an assigned task. Generally, once the program is completed, the organization is disbanded.

Program management is based on the team concept wherein the team consists of program functional managers and the program or project manager, as seen in Figure 4-3. The project manager is responsible for the overall success of the program. He or she is concerned with the planning and coordination of the program, rather than the details of execution for which the program functional managers are responsible. Project managers should be appointed in both the user and supplier organizations and are generally the "point men" for all activities. They are backed by functional managers, management, and many individual contributors in each of their respective organizations.

The project manager:

1. Receives authority from his or her management.
2. Is appointed by the appropriate management person in both the user and supplier organizations. He or she reports program progress as required to management, who reports upward in each organization as required.
3. Directs and controls functional areas in achieving program plans with respect to cost, performance, and delivery.
4. Is responsible for *what* needs to be done and *when* it must be done. Functional managers are responsible for *how* it will be done.
5. Is responsible for the company's commitment to specifications and requirements as set forth in the purchase order.
6. Is the principal go-between for his or her organization. This includes all program planning, program changes, reports, schedules, costs, delivery, communication, and mediating disagreements.
7. Has a principal role to direct, support, and assist the program functional managers in achieving their goals.
8. Calls regular project team meetings, chairs these meetings, and documents and distributes minutes and schedule changes.
9. Monitors and evaluates results, compares results with benchmarks, and initiates action to overcome problems and/or delays.

Functional managers represent each functional department required to participate in the project. They report to the project manager for program direction and to their function head for policy direction. They must have the authority to speak for their function. Their service on the project team may be full or part time, and they are responsible for the success of the function they represent.

Functional managers:

1. Are assigned from line functional departments to the project. They receive direction for what is required and when it must be complete from the program manager.
2. Receive direction for "how to accomplish the task" (policy) from their functional superior.
3. Must be given authority to make commitments to the program manager for the program on behalf of the functional superior and his or her organization.

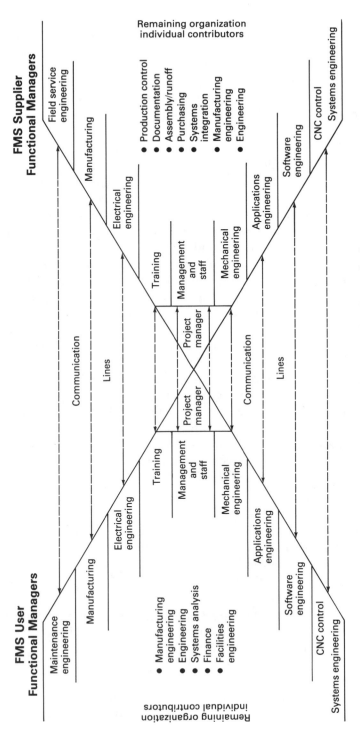

Figure 4-3 FMS user–supplier functional organizations.

They are responsible for all program-related functions pertaining to their department.

4. Have authority to issue direction regarding project matters to all people in the functional organization assigned to the project.

5. Inform both the program manager and their functional department heads as conflicts develop in order to resolve the conflict and reestablish project direction and schedule.

Although other functional organization approaches may work, the guidelines just outlined have been proved in many instances and are typical of FMS project management assignments. Communication lines between user and supplier functional managers, as seen in Figure 4-3, need to be established and remain open. However, copies of all paperwork should flow to or through the project manager to maintain overall project control and coordination. He or she is the focal point of the entire effort and must constantly be kept informed of project developments and proceedings.

Once functional organizations are in place, determining who (which organization) should be responsible for what should begin between user and supplier. It takes teamwork and a true partnership arrangement between user and supplier to make the FMS project win. A simple user–supplier responsibility chart similar to that depicted in Figure 4-4 is a good starting point. This helps clarify goals, objectives, expectations, and responsibilities early in the process and helps to avoid embarrassing and costly assumptions later.

SUPPLIER SELECTION

FMS vendors (suppliers), experienced in providing FMS systems and components, welcome clearly defined goals, objectives, and performance specifications by the purchaser to quote against. These provide the basis for developing system solutions especially suited for the user and saving time for both vendor and purchaser.

When selecting vendors, the main risk is that the vendor may have underestimated the task and the purchaser may not realize this fact. Well-prepared and detailed technical specifications by the user project team can help reduce this risk. It is unlikely, however, that an established builder of major CNC machines will underestimate or fail to meet a well-written specification. It is more likely that a subvendor (fixture or software supplier, for example) will misrepresent the specifications or overestimate his own ability. Consequently, any system vendor or subvendor can destroy a well-prepared timetable and implementation plan.

These views are not meant to be pessimistic, but realistic in nature. They reinforce the fact that a well-formed project team and strict attention to detail will help overcome foreseen and unforeseen project hurdles and roadblocks. In general, supplier selection guidelines should be:

FMS USER/SUPPLIER RESPONSIBILITIES

User Responsibility	*Supplier Responsibility*	*User and Supplier Responsibility*
□ Secure and maintain solid management commitment	□ Part processing and support equipment procurement/ manufacturing/ assembly	□ System physical planning and specifications
□ Upfront preplanning		□ System configuration and layout
□ Establish system goals and objectives	□ System quality assurance	□ Acceptance testing and implementation
□ Facility preparation	□ System integration	□ Training plan preparation
□ Part design, review, and analysis	□ Single-source hardware/software delivery, specifications, and support	□ Training implementation
□ Part selection, mix, and volume		
□ Organization/operation issues	□ Total system installation and operation	
□ User project team selection	□ Supplier project team selection	
□ System staffing requirements and selection	□ Subvendor selection	
□ Funding/justification	□ Overall training coordination	
□ Part programming	□ Documentation	
□ Tooling and fixturing requirements and specifications	□ Fixture design, manufacture, and build	
□ Redistribution of production to minimize disturbances; construction through installation and system acceptance		

Figure 4-4   An FMS user/supplier chart should be drawn up early in the planning stages to avoid misunderstandings and disagreements.

1. Prepare a Good Request for Quote (RFQ)

Adequate preparation and homework will help simplify this process and prevent errors of commission and omission. Specifics of machinery should be avoided in the RFQ. Think in terms of requirements and performance desired (part size, weight, mix, volume, lot sizes, accuracies, throughput, and so on). Set the level of system complexity and flexibility both desired and required. Infinite flexibility yields infinite costs. System boundaries, interfaces, and interconnectivity to other systems should be specified. Include plant limitations such as system area size, floor space allocation, temperature, overhead or underground limitations, and expected completion date.

2. Limit the Total Number of RFQs

Limit the number of vendors quoting on the system to three or four. The project team will have their hands full trying to technically, financially, and objectively evaluate these proposals as they arrive. If more than three to four vendors need to be considered, talk more with vendors and make additional visits to prior vendor installations to assess capabilities. Selecting only three to four vendors to quote will save user and supplier time, effort, and unnecessary work and anxiety.

3. Establish a Defined Acceptance Criteria

Acceptance criteria should be well defined in terms of meeting performance criteria and requirements. The FMS system, once installed and operational, should be tested in a complete and fully operational production run prior to acceptance. A detailed test plan, including unit or individual component testing, should be discussed with the selected supplier.

4. Prepare Reasonable but Sound Evaluation Criteria

Extra attention, preparation, and homework here can help to quickly select the right supplier. A rating system should be developed that adequately reflects the company's goals, objectives, values, and performance requirements. Technical criteria should include factors relating to safety, user friendliness, maintainability, and manageability, as well as compatibility with existing computer systems and operations. Supplier capabilities should be evaluated in terms of staff, past performance, and willingness to work with the buying team. Look closely at the potential supplier's commitment to automation as this is very important. Does the company apply what they sell in their own plants? Visit supplier installation sites and ask other users a lot of questions before final selection. Prepare questions ahead of time and build this into the evaluation criteria.

5. Establish Negotiating Posture

Hard-nosed, unbending positions regarding terms and conditions can deadlock negotiations and add considerable time to the supplier selection process. The approach should be one of reasonableness that promotes open discourse, concurrent engineering, and teamwork. This is the successful formula for flexible cells and systems.

SYSTEM DESCRIPTION AND SIZING

How big an FMS is needed is a function of the parts to be produced within the system: type, size, family, lot sizes, annual requirements, and so on. Detailed engineering studies and close analysis of component parts to be manufactured must first be performed in order to determine system size, description, and configuration. Further engineering and production studies will determine the tooling and fixturing requirements, inspection equipment, material, and ancillary support equipment. Additionally, results of detail part engineering studies will affect overall system layout and be determined by the available floor space and location, overhead and underground utilities, crane support, and material flow and movement patterns.

When considering component parts for FMS processing, the following factors affect system description and sizing and should be considered:

1. Part Volume and Variety

Part volume and variety will determine the basic design of the system. Identify the total number and mix of parts to be produced in the system. Can the parts be grouped into families with wide part variation existing between families, but limited variation existing within families?

2. Design Stability

Design stability of workpieces to be machined is extremely important. Later engineering changes can be easily accommodated if they are minor and affect only sections of part programs and cutting tools. Changes requiring part redesign and thereby requiring fixture redesign or modification are major and costly and should be avoided if possible. Current and future part design change considerations and stability should be reviewed and frozen at this early stage with engineering. Additionally, the user FMS project team and its engineering management should jointly develop design guidelines so future parts "fit the system." At the same time, disciplined engineering procedures should be developed and communicated to help secure stable part designs.

3. Cutting Tool Proliferation

In some cases it may be highly advantageous or even necessary to redesign parts specifically for FMS in order to reduce the large number of cutting tools required to produce the part. Redesigning parts to reduce a large cutting tool buildup can reduce part cycle times, tool inventories, and tool change time. This will also reduce the system size and cost, while increasing efficiency and performance. Reducing the large assortment of drills and taps, for example, through forced tool elimination is not an easy or short-term task. However, it is a task that can have a tremendous impact on system size, cost, and productivity, and it should be done early in the planning process.

4. Part Size Characteristics

Close attention should be paid to part size, weight, material type, dimensions, tolerances, fixturing requirements, cycle time, number of tools required to produce the part, and ease of loading and unloading. Some parts may be small enough to allow several to be fixtured on a pallet at one time (Figure 4-5), while others may be large and unwieldy and permit only one or two parts to be fixtured on a pallet (Figure 4-6). These factors will affect the type and size of the system and should be closely studied and evaluated. Additionally, wide variations may exist among parts selected to run through the system. For example, 85 percent of the parts selected for FMS processing may be aluminum and the other 15 percent steel. This is a wide characteristic variation and should be closely evaluated as to whether to accommodate the 15 percent steel parts within the FMS, thereby adding size, complexity, and cost to the system, or to have the 15 percent steel parts processed in some other manner. Future product and process growth considerations would also be key influencing factors.

5. Quantities and Lot Sizes

Current and future annual part requirements need to be closely studied. How part quantity and lot size requirements are released today is probably not consistent with economic order quantities (EOQ) in a fully implemented FMS environment. FMS increases part throughput due to increased spindle utilization with part setups, loading, and unloading being performed off-line while machine tools con-

Figure 4-5 Several small parts can be held (fixtured) on a pallet at the same time to reduce part handling time and increase machining time. (Courtesy of Remington Arms)

Figure 4-6 When parts are large and unwieldy, they are generally held or fixtured on pallets one or two at a time. (Courtesy of Remington Arms)

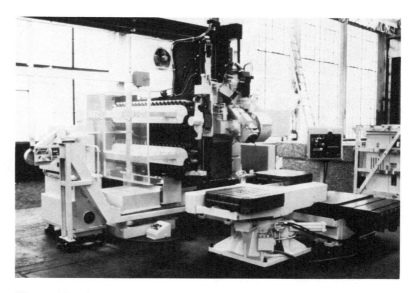

Figure 4-7 A ninety-tool magazine can accommodate a wide variety of cutting tools necessary for machining a part. (Courtesy of Remington Arms)

tinue to cut parts. Therefore, lot size and inventory level requirements can be reduced considerably from previous levels. This aspect needs to be closely studied and evaluated in order to determine system size. Another aspect of quantity and lot size review that affects system type and size has to do with load balancing. Work content and load (cycle times again) must be studied and balanced carefully among machining stations so that no machine is waiting for work for very long. This involves a further study of tooling and fixturing requirements in addition to quantities, lot sizes, and production schedules.

6. Tooling and Fixturing

To accommodate the increased part throughput and reduced lot size and inventory levels, redundant cutting tools and fixtures may be required. Close consideration must be given to cutting tool and fixture requirements to maintain throughput and balance the part processing load. Will additional cutting tools be added to the individual tool matrices to handle a variety of machining requirements? Will individual tool magazines (Figure 4-7) have enough available tool pockets to accommodate the extra tools? Will tools be shared and moved from machine to machine by means of an FMS-controlled tool delivery system (Figure 4-8) in order to hold down tool costs? Are duplicate fixtures required to accommodate production requirements? Answers to these questions will also affect the overall system description and sizing.

7. Manual Operations

Are any high-labor-content operations expected to be processed either internally or externally to the system? A high-labor-content operation, for example,

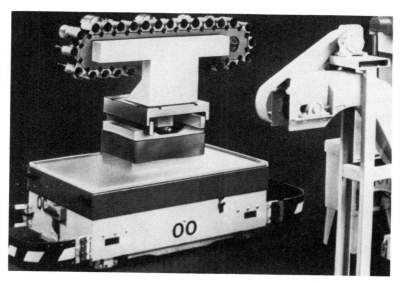

Figure 4-8 An FMS-controlled tool delivery system helps to reduce the number of cutting tools required, which in turn reduces the part cost. (Courtesy of Remington Arms)

might be extensive deburring or hand-tapping required because of part design, complexity, shape, or material type. The question becomes one of trying to automate the process through a special application or part redesign, thereby possibly affecting system description, size, and cost, or continuing with the manual process either internal or external to the system.

Other factors become important to system description and sizing, such as the number of pallets (Figure 4-9) for parts to be mounted on and processed through the system. However, the number of pallets is an indirect factor driven mainly by production requirements, part specifications, and how many parts, for example, can be mounted to a tombstone fixture (Figure 4-10). Nevertheless, all component part features, processing requirements, projected volumes, and selection of appropriate system operating characteristics need to be closely studied and analyzed in order to make an intelligent determination relative to overall system description, composition, and size. Detailed part analysis is the most important factor in deciding how big the system will be and what it will do. It is an analysis that involves the skills, cooperation, and disciplines of many individuals on both the user and supplier project teams.

SYSTEM DEFINITION AND SPECIFICATION

Once a detailed part analysis has been made and initial requirements have been worked out, the next step is to determine what the full FMS composition is. How detail parts are to be processed, and what kind of machine tools and how many are

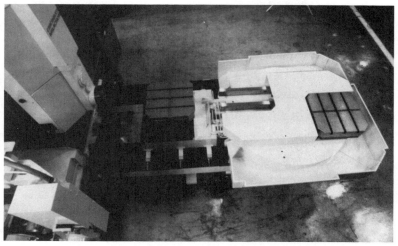

Figure 4-9   Pallets are used to hold (fixture) the part and move them through the machining operation and the FMS system. (Courtesy of Cincinnati Milacron)

Figure 4-10   A tombstone fixture allows many parts to be held (fixtured) for machining operations. (Courtesy of Remington Arms)

required, are functions of the part analysis. In addition, ancillary support equipment definition, such as for wash stations (Figure 4-11) and coordinate measuring machines (Figure 4-12), is also a function of the detail part analysis and overall operating specifications.

Figure 4-11   A wash station, part of the FMS, is used to remove chips, oil, grease, and the like, from the parts produced within the system. (Courtesy of Cincinnati Milacron)

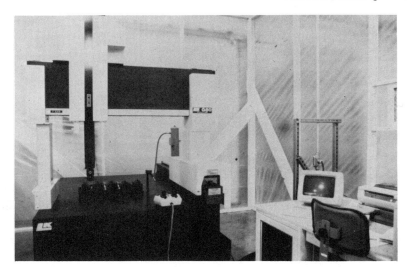

Figure 4-12   Coordinate measuring machines inspect parts produced on FMS machine tools to ensure that parts are machined to the proper tolerances. (Courtesy of Remington Arms)

Although many types of machine tools can be used in a FMS, the typical part producing workhorses for prismatic part FMSs are horizontal CNC machining centers. Two examples of these are shown in Figures 4-13 and 4-14. Horizontal CNC machining centers are ideal for prismatic part machining applications primarily because of their rotary index tables. Index tables on horizontal CNC machining centers permit machining on all sides of a workpiece in one setting, thereby saving time and maximizing machining efficiency.

Machining centers should be able to operate independently of FMS control if necessary. The machine's stand-alone capability should be a high priority if the machine is purchased as part of a time-phased plan of FMS installation. It may be required to make parts and begin paying for itself before the next installation phase of the plan is implemented. To run independently the CNC unit should be equipped with the capability to read either punched tape or floppy disk. If the tape or floppy disk reader were eliminated, alternative stand-alone capability of the machine would be lost. Multiple-part program storage is another valuable feature that would keep the machine running in case of host computer downtime. How much intelligence to leave at the machine is a matter of flexibility and insurance. An FMS is designed to operate automatically, but planning for "just in case" situations is playing it safe and smart.

One of the main issues affecting system definition and specification has to do with whether the system will be attended, lightly attended, or unattended. If lightly attended or unattended, certain features are required to handle critical tasks that were previously done by the operator on a conventional stand-alone machine. These features include broken tool detection and recovery methods, probe for locating and offsetting the machine to work surfaces, redundant tool selection for

Figure 4-13 Horizontal machining centers are the typical workhorses of an FMS. (Courtesy of Remington Arms)

Figure 4-14 Various types of horizontal machining centers are available to suit the size and shape of workpieces or the machining operations required. (Courtesy of Cincinnati Milacron)

Figure 4-15 Automated guided vehicles (AGVs) are used to move parts between various machines or stations within an FMS. (Courtesy of Remington Arms)

Figure 4-16 Queuing carrousels are storage areas (parking lots) for in-process or finished parts. (Courtesy of Remington Arms)

time-life expired tools, torque-controlled machining to protect tools and machines, high-pressure coolant control, and diagnostics reporting capabilities to the FMS host computer for both machine and control.

Every FMS system is different because each is based on the particular user's parts, production requirements, and overall operating needs and characteristics. Part cycle and load times will determine how many parts per hour can be processed through the system and the method of material movement. Long cycle times, for example, permit automatic guided vehicle (AGV) transporting of parts (Figure 4-15) and affect queuing carrousel size (how big the part parking lot has to be) (Figure 4-16). Short cycle times and higher production volumes, on the other hand, will require a faster material movement system, such as a power-roller conveyer system (Figure 4-17).

It becomes necessary at this stage to determine the production philosophy. Will parts be moved through the system in either a batch or ship-set philosophy? A batch philosophy (Figure 4-18) means that a batch (lot) of multiple identical parts

Figure 4-17   **A power roller conveyor system is used to transport parts where there is a short cycle time and high production volume. (Courtesy of Jervis B. Webb Co.)**

Figure 4-18   Identical parts being processed through an FMS system as a batch operation. (Courtesy of Remington Arms)

Figure 4-19   The ship-set philosophy produces one part from each of the defined part families. (Courtesy of Cincinnati Milacron)

will be produced for each part type of a particular unit or subassembly. A ship-set philosophy (Figure 4-19) means that all different parts that make up a particular unit or subassembly will be machined as a family or *set*. At the end of one production period, there will be one part of each type to *ship* and assemble as one complete unit or subassembly (ship set). Advantages of batch versus ship-set operating methodology can be determined through further production analysis with the system supplier.

Static analysis can then determine the minimum number of system components required to meet the stated production goals in a static or nonmoving state. Dynamic modeling techniques (systems in motion) are accomplished through computer simulation analysis and help to finalize system size, description, and definition. A general operational overview also helps determine final system definition and specifications. A simplified flow chart that omits detailed events is shown in Figure 4-20. Further discussion and analysis through a user–supplier partnership approach will bring system definition and specification to its final stage. Other important factors to consider relative to system definition and specification include:

1. For an FMS to operate unattended or lightly attended for a substantial part of a 24-hour day, every element in the system must be extremely reliable and have a long service life.

2. To minimize the possibility of jams, snarled chips, broken tools, and excessive tool inventories, parts should be programmed and the system should be operated at conservative speeds and feeds with lighter than usual depths of cut.

3. System software must have error-log reporting capabilities for capturing error signals and problems during unattended operation. The data can then be displayed upon request for action to be taken when people return to the system.

4. A reliable and possibly a central chip removal and coolant system that works unmanned will be required to avoid unnecessary machine shutdown.

5. Techniques must be preplanned for the system to automatically solve any problems occuring during unattended operation to prevent machine or entire system shutdown.

6. Enough workpieces must be in the system to keep machines fed during the unattended mode.

7. There must be enough tools and redundant tools in each machine's tool storage matrix to meet part requirements and variety during unattended operation and to replace broken or worn tools when necessary.

8. System software must have look-ahead capability to make a short-run forecast of upcoming work orders, tooling, and part program needs. Without an automatic tool delivery system, constraints will need to be placed on the system as parts can be routed to only those machines preloaded with the needed tooling.

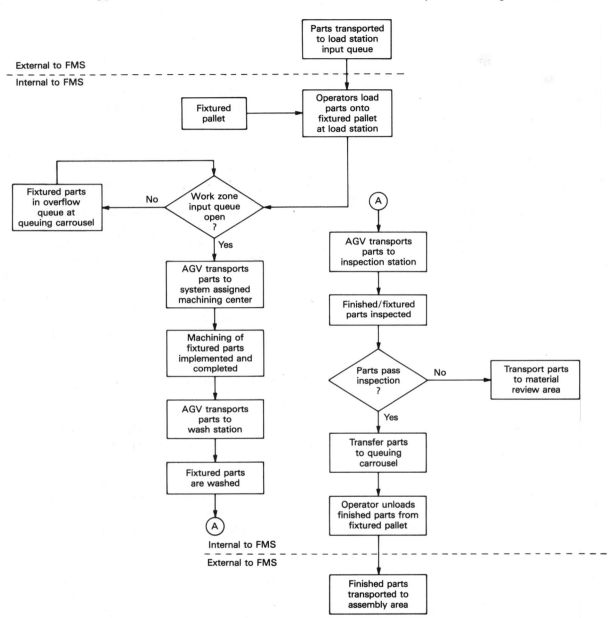

Figure 4-20   Part and fixtured pallet flow.

GENERAL LAYOUT AND CONFIGURATION

FMS layout and configuration are mainly determined through prior system defini-tion, description, and sizing analysis. Although other detailed elements are in-volved, the principal factors influencing system layout and shape are:

1. Type of parts to be processed
2. Production requirements
3. Type and time of operations required
4. Number and types of machines
5. Type of material transport selected
6. Available floor space, size, and shape
7. Proximity and convenience to utilities, support services, and incoming and outgoing material

Generally, FMS layouts can be classified under four basic types or categories:

1. Progressive or in-line (Figure 4-21)
2. Closed loop (Figure 4-22)
3. Ladder (Figure 4-23)
4. Open field (Figure 4-24)

The progressive or in-line and closed-loop types shown in Figures 4-21 and 4-22 are more dedicated FMSs or cells because they:

1. Handle a relatively small and dedicated family of parts
2. Cover medium to very large part size and configuration
3. Tend to have long cycle (in-cut) times
4. Generally cover a greater number and variety of processing operations

These characteristics allow distribution of total part processing time over more machines. However, some machining specialization can be accommodated among processing machines of the same type.

The closed-loop concept is often used where machines are arranged in a progressive, general order-of-use loop. The computer can route parts through the system having operations performed in a sequential manner at some machines while avoiding others. Loop-type systems can also be straightened out to form an in-line arrangement for progressive pallet movement down the line or expanded to include side paths or steps (Figure 4-23) as part of a ladder type of layout.

The open-field FMS (Figure 4-24) is most commonly used where machines of the same type are more closely arranged and where component parts can be easily

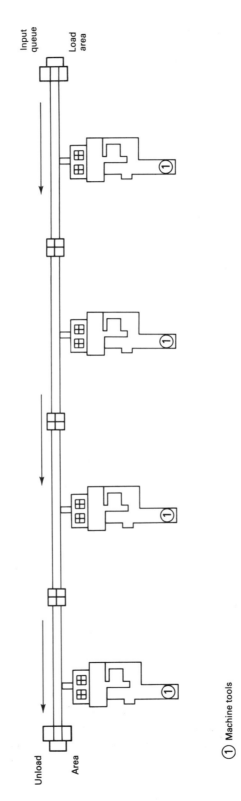

Figure 4-21 A progressive or in-line FMS system moves work from the load area through each machining station and then to the unload area.

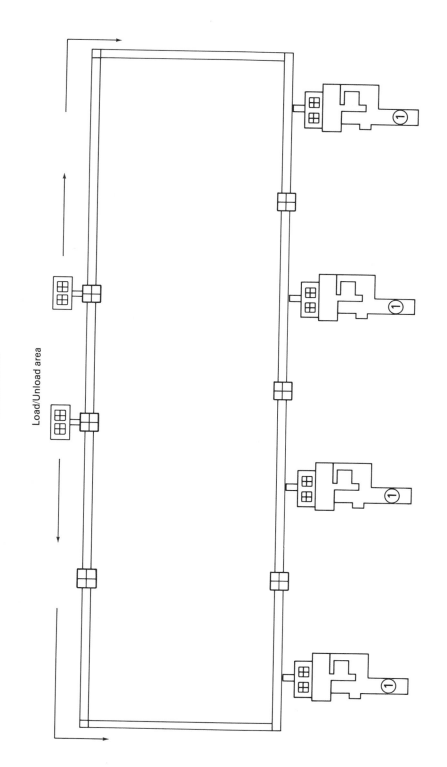

Figure 4-22 In a closed-loop FMS system, parts are moved around a conveyor loop from machine to machine.

FMS computer room

Load/Unload area

① Machine tools

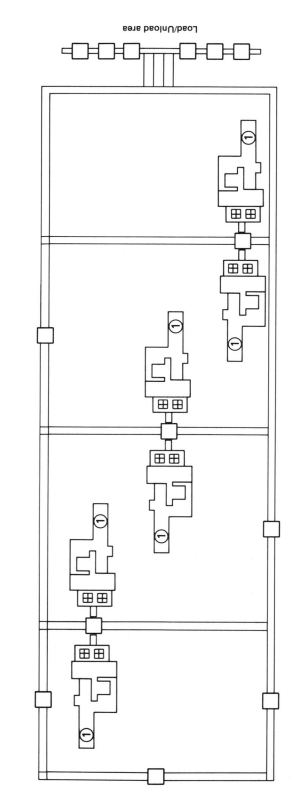

Figure 4-23 In a ladder-type FMS system, the machine tools are located on the steps of the ladder, and the parts can be routed to any machine tool or machine group. This allows two machines to work on one pallet at the same time.

Load/Unload area

FMS computer room

① Machine tools

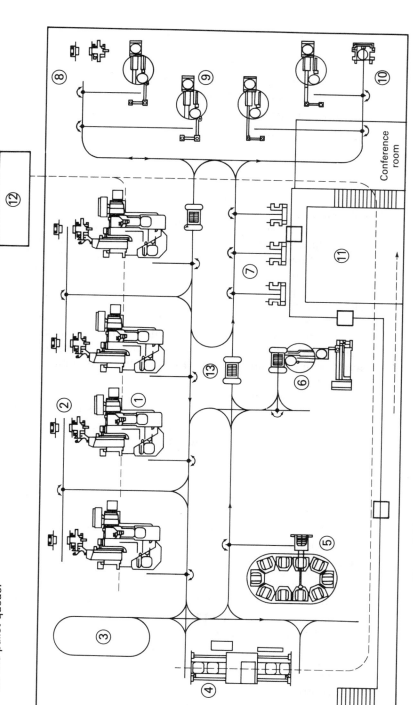

① Four CNC machining centers.

② Four tool interchange stations, one per machine, for tool-storage chain delivery via computer-controlled cart.

③ Cart maintenance station. Coolant monitoring and maintenance area.

④ Parts wash station, automatic handling.

⑤ Automatic work changer (10 pallets) for on-line pallet queue.

⑥ One inspection module: horizontal type coordinate measuring machine.

⑦ Three queue stations for tool delivery chains.

⑧ Tool delivery chain load/unload station.

⑨ Four part load/unload stations.

⑩ Pallet/fixture build station.

⑪ Control center, computer room (elevated).

⑫ Centralized chip coolant collection/recovery system (----- flume path).

⑬ Three computer-controlled carts with wire-guided path.
Cart turnaround station (up to 360° around its own axis).

Figure 4-24   In an open-field system, machine tools and support equipment can be arranged in any order to suit the the needs of the user.

sent via AGV to any available machine. Factors influencing this type of arrangement are:

1. Numerous parts per family (wide variety)
2. Small to average part size
3. Short part processing cycle times per average part
4. Small number and variety of operations per average part

These particular characteristics tend to minimize the different kinds of machine tools, but increase the system cost because of the sophisticated software required to manage and control a wide variety of parts and resources in a dynamic, flexible environment.

One area that deserves special attention relative to FMS layout has to do with material flow patterns. Often considerable effort is spent arranging equipment to optimize efficiency and part throughput internal to an FMS, with very little effort spent on optimizing material and part throughput flow patterns external to the FMS. Consequently, even though the internal FMS operations may be very efficient and productive, incoming raw material and outgoing completed parts may be a bottleneck area, all because the system was layed out inefficiently relative to the overall material flow pattern. Material flow and movement patterns must be given careful consideration during system layout and configuration analysis in order to maximize internal and external FMS efficiency and part throughput.

System configuration, although a function of the system sizing and operational scenario discussed earlier, is a detailed study of the individual FMS elements and components and how each can best be arranged to maximize FMS productivity and performance. Through system description, sizing, and definition analysis, an equipment list is prepared similar to the following:

5	horizontal five-axis CNC machining centers
2	ten-position pallet queuing carrousels with two load–unload fixture build stations
1	wire-guided transportation system (three-AGVs) with cart maintenance station
1	calibration cube for automated checking of machine alignments
2	material review stands
1	central coolant and chip recovery system
26	pallets for system part transportation
1	FMS computer control system
1	wash station for part and pallet cleaning
1	tool gage for operator setting and gaging tool lengths
1	coordinate measuring machine (CMM)
1	tool-build module for building up fixtures prior to system entry

Once a complete equipment list has been developed, the following general factors should be considered when arranging an FMS:

1. The FMS computer control center–system manager's office should be elevated, as seen in Figure 4-25, to provide a range of vision over the entire system. In some cases, colored, highly visible lights are wired into each processing station to indicate equipment status. For example, (a) yellow, machine in cycle (FMS mode), (b) blue, machine in automatic wait (waiting pallet delivery mode), (c) white, machine in cycle but waiting due to NC program stop mode, (d) green, machine in manual mode, (e) red, machine in failure mode.

2. Incoming and outgoing material flow patterns should be as efficient as possible.

3. Are backup AGVs required? If so, are they located in an area where they are ready for duty and service such as battery exchange, recharge, and maintenance of facility AGVs?

4. Are power panels conveniently located for accessibility and serviceability?

5. If a central coolant–chip recovery system is used, pumps, hydraulics, motors, and controls should be suitably protected and easily accessible for standard maintenance and service. Nonoxidizing materials should be used for coolant plumbing.

Figure 4-25   An elevated computer–manager room allows a full range of vision of the entire FMS system. (Courtesy of Remington Arms)

Figure 4-26 **The tilt mechanism can be added to the load–unload station to allow for easy loading and unloading of large, unwieldy parts. (Courtesy of Cincinnati Milacron)**

6. Depending on part tolerances to be held by FMS processing equipment, and subsequently checked by inspection equipment, opening and closing outside doors, for example, can cause temperature fluctuations, thereby affecting part tolerances. If the facility is not environmentally controlled, processing and inspection machines should be placed away from opening and closing doors, cold air returns, and the like, to provide a more stable environment.

7. If parts are large, cumbersome, and unwieldy, a tilt mechanism similar to that depicted in Figure 4-26 should be placed into the system at the load–unload station. A tilt mechanism allows operator ease of part loading, unloading, and clamping because the tombstone fixture (Figure 4-27) can be rotated 90 degrees from the vertical to horizontal position and subsequently rotated 90 degrees back to its original position for reentry into the system.

FACILITY PREPARATION PLANNING

Facility preparation and analysis is another area of the overall FMS planning process involving close attention to detail. Often the services of outside civil engineering and construction firms must be obtained through a contract bid and award process in order to prepare the site for the FMS.

Once a reputable civil engineering and construction firm has been retained, a three-part alliance is now basically formed between the FMS user, supplier, and construction–engineering firm. Although site preparation planning and construction are mainly the responsibility of the FMS user, a close working relationship must be formed between the appropriate members of each project team (FMS user and supplier) and the construction superintendent in order to focus on the concerns

Figure 4-27 Tombstone fixtures can be rotated 90 degrees from a vertical to a horizontal position for easier loading and unloading of parts. (Courtesy of Cincinnati Milacron)

of each for overall project success. Open communication and friendly discourse must be maintained to achieve the level of cooperation required to successfully complete a project of this magnitude.

Site preparation studies begin with the user deciding where the FMS should be located.

1. Should the FMS be located in an existing or a new building?
2. If the FMS is to be located in an existing facility, how much will the present production be disrupted as the facility is vacated in preparation for the FMS installation?
3. How will the FMS layout fit within the planned facility and affect material flow movement, accessibility, and maintenance concerns?
4. Should the layout be changed to suit the facility or the facility changed to suit an efficient FMS layout?

Questions such as these and many others need to be resolved early in the planning process for the FMS facility.

One big issue relative to site preparation that must be decided very early is who (user or supplier) is responsible for what. Generally, as part of the FMS proposal and specification process, site preparation responsibilities are clearly defined and agreed on. The site preparation general responsibilities chart shown in Figure 4-28 lists some of the more common examples of FMS user and supplier

| SITE PREPARATION GENERAL RESPONSIBILITIES | |
User	*Supplier*
SYSTEM LAYOUT — System layout approval	SYSTEM LAYOUT — System layout engineering
FOUNDATION — Civil engineering responsibilities — Surveying (above and below ground requirements) — Soil samples — State and local permits	FOUNDATION — Minimum requirements and specification for foundation
CONSTRUCTION — Contract bid/award — Excavation — Concrete and floor reinforcement — Building (if required)	CONSTRUCTION — General specifications and requirements
UTILITIES — Electrical power supply, air, water, drains — Provide source material (wire, conduits, piping, etc.) and installation labor for both above and below grade — Decide what is above grade and what is below grade —Define source locations — Engineer above grade routing — Supervise above grade installation	UTILITIES — Review user's above grade engineering — General specifications, requirements, and power drop locations — Define general routing above grade — Engineer below grade routing — Supervise below grade installation

Figure 4-28   A site preparation responsibility chart clearly outlines both user and supplier responsibilities.

SITE PREPARATION GENERAL RESPONSIBILITIES

User	*Supplier*
COMMUNICATION WIRING – Engineer above grade routing – Provide all material except wire (conduit, junction boxes, etc.) – Labor and supervision for above and below grade installation	**COMMUNICATION WIRING** – Specify requirements – Define general routing above grade – Specify labor skills required – Provide wire
EQUIPMENT – Excavation/installation central coolant and chip recovery system – Provide labor and tools – Rigging – Provide enclosures for environmentally sensitive equipment (CMMs)	**EQUIPMENT** – Provide expertise to closely supervise setup, reassembly, hookup, and runoff to OEM specifications
COMPUTER ROOM – Procure and install enclosure	**COMPUTER ROOM** – General specifications and requirements – Install computer equipment
MATERIAL-HANDLING SYSTEM – Engineering, supervision, material, and labor for battery charge station (extra)	**MATERIAL-HANDLING SYSTEM** – Engineering, supervision, material, and labor for floor cutting, floor wiring, installation, and runoff
SAFETY – Safety audits pre- and postinstalled system – Perimeter guarding erection – Enforcement of safety guidelines and procedures	**SAFETY** – Equipment – Safety audit of installed system; pre-user acceptance – General specifications; perimeter guarding

Figure 4-28   (*Continued*)

Figure 4-29   The first step in actual site preparation for an FMS is excavation. (Courtesy of Cincinnati Milacron)

responsibilities. This is not necessarily indicative of how all site preparation responsibilities should be broken down, as these elements in the planning process are also subject to negotiation. Construction of a new facility can, in some cases, be easier and perhaps less costly and disruptive than renovating an existing facility. Some factors that could affect this decision are:

1. Structural condition of the existing building
2. Thickness and condition of the existing floor
3. Heating, ventilating, and environmental control of the building
4. Underground sewer lines, drains, and other utilities
5. Power requirements
6. New versus old material flow patterns
7. Existing versus required overhead crane support
8. In-ground central coolant flume path requirements

Floor preparation is an important area of consideration as AGVs require a smooth, level surface for operation. And individual processing equipment requires the appropriate foundation thickness, underground reinforcement, and antivibration mounts, if necessary.

Excavation (Figure 4-29) is usually the first step in the actual site preparation.

Figure 4-30   Deep excavation is required inside the building for the flume system. (Courtesy of Cincinnati Milacron)

Figure 4-31   Excavation is required outside the building in some cases for the central coolant tank. (Courtesy of Remington Arms)

Figure 4-32   Provision must be made for all services required for an FMS, including air, water, and drain locations. (Courtesy of Remington Arms)

If an existing facility is being renovated, the existing floor may require complete or partial removal, depending on floor specifications for the new equipment. Excavation is also the first step for a new site, including underground preparation. If a central coolant and chip recovery system is to be installed, deep excavation (Figure 4-30) would be required within the building, while deep excavation outside and adjacent to the building is required for the central coolant and chip recovery tank (Figure 4-31).

Other important areas of consideration for site preparation include electrical power, air, water, drain locations, and overhead crane support. These should be given careful consideration (Figure 4-32) relative to the overall FMS layout, with particular attention paid to what is above and below grade.

System expansion should be considered and planned for early in the planning process. If system expansion is anticipated, available floor space for additional equipment, material flow, and tool and fixture storage should be planned and allocated at this stage. Power drop locations, other utilities, and building entry and exit points should be discussed and considered relative to site preparation for future expansion.

SUMMARY

1. Sound planning is the single most important factor in successful automated manufacturing.

2. Twenty-five percent of the gain expected from automation projects such as FMS can be achieved by adopting a proper plan.

3. The project master plan is *the* controlling document for the entire project that establishes commitments, responsibilities, deadlines, and activities. It must be respected and closely followed.

4. Asking questions at the right time (planning process) is as important as asking the right questions.

5. Program management is based on the team concept, whereby the team consists of program functional managers and the program or project manager.

6. Teamwork, communication, and partnership are the key ingredients between FMS user and supplier to make the project succeed.

7. FMS size and description are functions of the parts to be produced within the system. Detailed part analysis is the most important factor in deciding how big the system will be and what it will do.

8. Material flow patterns must be given careful consideration during system layout and configuration analysis in order to maximize internal and external FMS efficiency and part throughput.

9. Detailed and well-prepared technical specifications by the FMS user project team are the best way to minimize the risk of the supplier underestimating the task.

10. FMS layouts can be classified in four basic types: (1) progressive or in-line, (2) closed loop, (3) ladder, and, (4) open field.

11. Overall system user–supplier responsibilities, including site preparation, need to be clearly defined and agreed on in the planning process.

REVIEW QUESTIONS

1. In general, planning works best when:
   (a) The individuals doing the work are accountable
   (b) The principal contributors collectively agree
   (c) All available people are put on the project
   (d) Only (a) and (b)
   (e) All of the above

2. The _____ is *the* controlling document for the entire FMS project.

3. It is not necessary for senior management to participate in the actual planning preparation process. True or False?

4. Program management is based on the team concept whereby the team consists of program functional managers and the _____ .

5. When selecting FMS vendors, the main risk is that the vendor may have underestimated the task without the purchaser realizing this fact. True or False?

6. How big an FMS is needed is a function of:
   (a) How many processing machines it will contain
   (b) How much floor space is available
   (c) The parts to be produced within the system (type, size lot sizes, and annual requirements)
   (d) The number of people available to attend the system

7. Part volume and variety will not determine the basic design of the system. True or False?

8. Explain the difference between batch and ship-set manufacturing.

9. Generally, FMS layouts are classified as (1) progressive or in-line, (2) closed loop, (3) ladder, and (4) _____ .

10. Material flow patterns, both internal and external to an FMS, are always improved through FMS installation. True or False?

11. FMS foundation preparation is the _____ responsibility.

12. Both system user and supplier have responsibility for overall physical planning and specifications. True or False?

5

Human Resources

OBJECTIVES

After studying this unit, you will be able to:

- ☐ Understand the importance of human resource considerations to a successful FMS installation.
- ☐ Identify some key individual attributes to question and rate during the FMS team interviewing and selection process.
- ☐ Discuss the value of teamwork, communication, and involvement to a successful automation project.
- ☐ Understand key issues surrounding changes in the supervisor's role.
- ☐ List some important factors to consider relative to FMS training.

INTRODUCTION

Organizing and staffing for automation require changes in the way a company organizes its plant and manages its people. Issues abound relative to retraining, redeployment, job security, loyalty, new job assignments, seniority, experience, and many others. How these issues are approached and handled relative to automation can have a serious impact on overall FMS success.

FMS equipment must obtain higher utilization rates in order to reduce costs, improve productivity, and increase profitability. Obtaining higher utilization rates means operating equipment in the range of fifteen to twenty shifts per week and during the unsocial hours of second and third shifts. Operating during these unsocial hours can be a strong argument for increased automation as it becomes more difficult to find qualified people to work in a conventional shop environment on second- or third-shift assignments.

Unquestionably, with FMS individual responsibilities will change. People's skills will broaden, but fewer jobs in the FMS arena will be required. Individuals must be more flexible, adaptable, and knowledgeable, performing tasks previously done by several trades and individuals. Such responsibility and skill upgrading places increased emphasis on the supervisor's role, employee training, and management attention to human resource concerns and considerations.

A successful people–automation connection is built on teamwork, communication, and employee involvement in an environment where openness, honesty, and sincerity prevail. Managing the elements of FMS change in such an environment provides growth, challenge, and opportunity for the company and its employees.

STAFFING CONSIDERATIONS

Generally, companies considering FMS or any type of major factory automation program give less attention to staffing issues than to technical issues. However, if a company is to be successful with advanced manufacturing technology, it may be necessary to change the way the company organizes its plant, staffs its operations, and manages its human resources.

How an FMS should be staffed depends on how it will be organized and what it will do. Will the FMS operate unattended or lightly attended? Will all the work content be internal to the FMS or will some be external? How many full shifts will the system be in operation? Questions like these and many others must first be answered in the planning phase of the project. Once the technical and operational issues are addressed, organizational and staffing can then be considered.

How to staff an FMS installation begins with a fundamental understanding of the functions required to operate and maintain the system, along with the skills required to perform those functions. Early identification of these functions and skills is an important first step in staffing an FMS.

General functions and system positions required to operate and maintain an FMS are:

1. *FMS system manager* (Figure 5-1): executes system commands to host computer that initiates action and activity of machine tools and related processing equipment. The system manager inputs scheduling and work-order requirements and is in charge of the total system operation.

2. *FMS system operator* (Figure 5-2): loads and unloads new and completed parts and related tooling into the system. He or she is responsible for part quality and accountability from part entry to part exit. These individuals need to know how to manually operate and interface with CNC machine tools, CMMs, and other part-processing and material-handling equipment in case manual operation is required as a result of host computer downtime problems.

3. *Mechanical–electronic maintenance technicians* (Figure 5-3): diagnose, troubleshoot, and repair machine tools, CNC units, CMMs, material-handling, and other part-processing equipment. They perform regular preventive maintenance activities and functions but do not troubleshoot or repair computers or programmable controllers.

These general system positions assume all part programming and tool design work has been completed prior to system operation. If new parts are to be added to the system at a later date, additional manufacturing engineering skills will be required to design fixtures and fulfill the new part programming requirements. Obvious questions regarding functions and skills of staff personnel revolve around how many people of what type are required to support the system and how they should be organized.

There are no specific answers relative to how many individuals of each type are needed to support an FMS. Variables such as how many shifts the system will be in continuous operation, system complexity and capacity level, and individual skills and how they will be organized, to name a few, all serve to make staffing considerations for each FMS a unique case. However, it can be seen that with FMS fewer individuals are required to support more highly utilized machine production

Figure 5-1   The FMS system manager executes system commands that initiate action and activity of equipment. (Courtesy of Kearney and Trecker Corp.)

Figure 5-2   The FMS operator is responsible for part quality from part entry into the system to part exit from the system. (Courtesy of Interlake Material Handling Division)

rates in a complex, changing environment. Consequently, the individuals support-ing these systems must be more flexible, trainable, and due-date driven in order to accommodate the increased production requirements associated with automated systems.

FMS requires more technically competent people because of its increased complexity and computerized dependencies. Additionally, the worker mind-set must change from that of dealing with individual tasks, assignments, equipment, and components to that of a more globalized approach to keeping the entire system in operation. Successful companies with FMS systems have selected individuals who have a whatever-it-takes-to-get-the-job-done attitude and are committed to keeping the system productive and in operation. This is extremely important, as any ''that's-not-my-job'' attitude can destroy the system and people continuity required to keep the FMS producing at optimum productivity levels.

Ensuring the right attitudes in the FMS staff is a function of selection, training, and organization. All these will be discussed in more detail later in this chapter. However, each should be mentioned here as a primary requirement to staffing an FMS.

Figure 5-3 Mechanical–electronic maintenance technicians diagnose, troubleshoot, and repair part processing and support equipment in addition to performing preventive maintenance functions. (Courtesy of Kearney and Trecker Corp.)

1. Selection

This is the most important element to be considered in staffing an FMS. Stringent job-related criteria elements should be developed and individual skills, abilities, and experiences should be evaluated against requirements in order to make the interviewing and selection process as quantitative and objective as possible. Favoritism, politics, and cronyism, although existing, have no place in the selection of qualified people for an FMS. Jobs may require internal posting before the hiring process actually occurs. In-house skills may not be available for all job requirements. Consequently, external hiring may be required to fill certain positions or for backups.

2. Training

Another vitally important element in staffing an FMS, training is often an afterthought and is not given the up-front consideration required for FMS success. Training considerations are the responsibility of both the FMS supplier and purchaser. Skill deficiencies must be recognized, identified and overcome by competency-based training activities and not left to catch-as-catch-can supplier training practices.

3. Organization

How the people will be organized can have a serious impact on FMS success and productivity levels. Will the people be organized in teams? Will jobs and shift assignments be rotated to promote more uniformity and increase the overall system knowledge base? Will individual assignments be nonrotational or semirotational in order to take maximum advantage of individual specialized skills and experiences? Most successful FMS systems are organized around the *team* concept of operation. Each team is responsible for overall system operation, safety, and maintenance. Individual assignments may or may not be rotated within each team as long as each team knows they are responsible for keeping the system operational and maintaining expected productivity levels.

Staffing for FMS or any factory automation project involves change that can reach deep into the operating and management ranks of the organization, precipitating effects that inevitably may bring about additional change. General staffing changes brought about as a result of implementing FMS or other factory automation projects consist of:

1. Stronger than usual partnerships between management and labor
2. A highly skilled, flexible, problem-solving and unified work force
3. Closer interdependencies among work activities, functions, tasks, and employees
4. Fewer employees in a team or unit responsible for a given product, part, or process
5. Higher capital investment and utilization level per employee
6. Broader ranging and more costly consequences for mistakes, malfunctions, and inattentiveness
7. Increased requirements for mental versus physical skills on the shop floor
8. A more technically involved, innovative, responsive, and risk-taking management team
9. Less "me" and more "we" orientation as a result of the team and system approach
10. Higher longevity and less attrition as a result of automation improvements and increased learning and job challenge

TEAMWORK, COMMUNICATION, AND INVOLVEMENT

Any factory automation project as large and broad in scope as FMS requires the cohesive blending of many skills, disciplines, talents, and abilities in order to be successful. How these human resource assets are organized and directed early in the planning process and how the people feel about such an undertaking can have a major impact on the project.

Early in the planning process, long before any equipment is ever on the shop floor, the process of teamwork, communication, and involvement should begin. Usually this begins with a series of management-held meetings for all employees both directly and indirectly affected by FMS. These meetings should be informal presentation first, questions and answers after type of sessions held to a maximum of fifteen to twenty people for each session. This is an ideal size group, as larger groups can provide questions that extend the meeting beyond the allowed time, and smaller groups may lack participation.

Management in these meetings must aim to make the communication process open, friendly, honest, and sincere. It is much easier to introduce new technology when management has established an open relationship with all employees. Specific details may not always be available because management may not yet be sure what action will be taken. However, keeping all employees up to date by reviewing plans and changes and listening to concerns will keep communication lines open between management and labor, verify or eliminate rumors and relative uncertainties, reduce anxieties, and help to motivate those interested in being a part of advanced manufacturing technology.

Employee meetings (Figure 5-4) should be informal get-togethers where communication flows in both directions. Superintendents–managers should give thought to being creative, like buying FMS T-shirts for the project team and using various forms of media such as video tapes about FMS to get the people enthusiastic. Also, scheduled supplier presentations relative to the FMS being purchased should be presented to employees. Presentations and discussions should focus on the quality and productivity gains of FMS, on "making jobs easier" and "working smarter," and what the company is doing to improve cost, quality, and delivery factors in the face of intensifying competition.

Meetings by themselves, however, do not mean that open communication has been accomplished. Meetings must contain enough useful information about FMS project and implementation plans to be meaningful. Without knowing enough about project details, it becomes difficult for employees to provide feedback. Consequently, they may consider management's communication attempt to be less than sincere, misinterpret the true intent of the meetings, and lose interest.

Communication is not something that just happens. It requires planning, effort, and work. The amount of communication effort required to convey project information is directly proportional to the size, scope, and complexity of the project itself. Large-scale factory automation projects like FMS require considerable time and effort in communicating plans and strategies, along with listening, evaluating, and acting on employee concerns and considerations. Employees want information, knowledge, recognition, and opportunity. Good communication provides the conduit to make it all happen.

Employee involvement really goes hand in hand with communication. People basically want to be involved, feel informed, and have some say about their jobs. Management can communicate desired values to employees and help them feel involved through a variety of ways other than group meetings, presentations, video

Figure 5-4   Employee meetings should be informal get-togethers where communication flows in both directions. (Courtesy of Cincinnati Milacron)

tapes, and encouraging feedback. Use of newspapers, bulletin boards, and special letters to employee's homes all help to convey the plant culture required to make FMS a success.

Management should model, by their own example, the values they wish to convey to the operating team. This means encouraging openness, rational decision making, and employee involvement and participation in the decision-making process. It means, in unionized plants, getting union officials involved in the planning process by asking them to visit vendors and make recommendations on equipment. Involvement also means trusting people and allowing them to develop and use their fullest capabilities at work. Existing team members should even be involved in developing the selection criteria, hiring practices, and work group procedures for new team members as the project evolves and progresses.

Getting people involved through a participative management style is generally easier with a new plant start-up than with an existing one because individual skills, experiences, abilities, and attitudes can be more closely scrutinized and a rigorous selection process adopted. Resistance and skepticism may run high in an existing plant with a more traditional, authoritarian management style, thereby impeding the entire teamwork effort. A plant's culture, however, can be changed, but benefits and risks are associated. Plants where cultural change has been adopted and accepted generally have a more committed and multiskilled work force, lower turnover, and higher return on their automation investment. On the

Figure 5-5 Employee involvement programs have become state of the art in human systems to maximize use of each employee's skill, knowledge, and experience. (Courtesy of Kennemetal, Inc.)

other hand, changing a plant's culture, especially an authoritarian one, can create suspicion, uncertainty, and instability for employees, while replacing predictability and order with disharmony, anxiety, and unclear expectations.

Employee involvement programs and participative management have been gaining acceptance on an international scale, particularly with FMS and other automation projects, as traditional organizational structures become ill suited to handle increasing and intensifying worldwide competition. Participative management and employee involvement programs (Figure 5-5) have become state of the art in human systems, just as FMS and factory automation have become state of the art in technical systems.

Teamwork is the human resource essence of advanced manufacturing systems. With a team approach, the goal is more important than the means, and traditional functional labels such as "machine operator" and "maintenance technician" should not disguise the increased interdependence of the functions. Decisions made by people in these functions were once relatively independent. With a team approach they must now be made jointly, usually with information that is shared or mutually generated. Organizations can facilitate such interdependence by equalizing status and pay and providing training classes and workshops for improving interpersonal and problem-solving skills. And traditional organization charts may be replaced by diagrams reflecting the consultative forms of decision making that increasingly characterize factory operations.

By organizing FMS responsibilities around teams, the scope of work is broadened in several ways:

1. Machine operation becomes more machine tending and maintenance becomes everyone's responsibility as simple troubleshooting, debugging, and problem diagnosis become everyone's priority in order to attain agreed on goals and maximize equipment utilization.

2. Routine machine maintenance becomes the team's responsibility also. This includes greasing and oiling machines, maintaining adequate coolant conditions, and performing housekeeping tasks.

3. Responsibility is increased for quality; in-process inspection and statistical process control, with emphasis on prevention of discrepancies instead of after-the-fact detection, is a team-measured goal.

4. Simple NC machine programming is learned by all team members (Figure 5-6) as periodic manual intervention is required in an FMS environment.

5. Cutting-tool nomenclature and how to detect tooling abnormalities is taught in order to make recommendations for improved cutting conditions, part program optimization, fixturing, or methods improvements.

6. Increased decision making is encouraged on matters concerning scheduling and machine utilization levels. The team is given the tools to do the job and is measured on its input, output, quality, and ability to meet expected deadlines.

The benefits of the team approach include:

1. *Multiskilling:* Team members learn from each other and build new skills in a team environment.
2. *Flexibility:* As individuals build new skills and learn from each other, they increase their value to themselves and to their companies, thereby enabling them to handle a variety of problems and situations in a high-tech operating environment.
3. *Leaner organizations:* Increased flexibility of the work force means less specialized skills will be required and organizations can run leaner as more skills are incorporated into fewer individuals.
4. *Job rotation:* Organizing in teams enables job rotation among team members, further promoting multiskilling and flexibility of the work force. Job rotation adds challenge to individual assignments and helps to keep employees interested and motivated.

Additional benefits of organizing around work teams are:

☐ Reduced operating costs
☐ Higher performance levels

Figure 5-6 Simple machine programming is learned by all team members as manual CNC intervention is sometimes required. (Courtesy of Remington Arms)

☐ More rapid problem diagnosis
☐ Increased worker responsiveness and feedback
☐ Improved quality through better trained people
☐ More knowledgeable and self-directed work force
☐ Reduced dependency on regulations and procedures
☐ Teams become self-auditing
☐ Reduced number of job classifications
☐ Increased element of trust and sharing
☐ More interdependency on each other
☐ Reduced employee turnover

Building teamwork, communication, and involvement can also have negative effects if not handled correctly and sincerely. Some possible side effects to guard against include:

1. Arrangement of work responsibilities that deskills, routinizes, or fragments work
2. Inadequate evaluation and wage payment systems
3. Increased job pressure on employees
4. Too much work responsibility moved from operating to managerial or professional personnel

5. Undesirable or unachievable work schedules and deadlines
6. Employee communication and social interaction is restricted, thereby tending to dehumanize the work place (system design, layout, and/or organization restricts people from talking together)
7. Personal privacy is invaded or aggravated to strain union–management relations (too much lack of trust and over-the-shoulder management checking)

In most cases, FMS and other factory automation teams can be formed through job postings where people can "sign up." Meetings can communicate to applicants the specific requirements and personal attributes required for selection. Selection of the operation team, just as with selection of the project team, should not be left to chance. Specific evaluation criteria, as will be discussed later, should be used to make the process as objective as possible.

FMS and other advanced technology applications place a premium on internal motivation, alertness, and problem solving and increasingly require intellectual instead of physical workers to maintain productivity.

Although not every worker wants the increased responsibility and challenge of working with an FMS, most welcome change, challenge, and the opportunity to participate in a state of the art manufacturing environment. Jobs in a team environment are usually open ended, allowing team members to assign and coordinate them. The central idea is to provide relatively few rules and regulations, leaving team members as much discretion as possible in pursuing the team's preestablished goals. Many companies, with installed FMS systems, for instance, have organized around three shifts of five-person teams each. Each individual is cross-trained in the essential elements of system operation, preventive maintenance, CNC programming, and troubleshooting. Supervision works to make attainment of the team's goals easier and to provide technical assistance.

The increased interdependence of activities and complexity of automated systems necessitates a teamwork structure to maximize productivity and equipment utilization. As mentioned earlier, state of the art human resource systems are required to keep pace with state of the art manufacturing systems. In general, human resource systems must be designed to:

1. Broadly define jobs to encompass more interrelated tasks
2. Effectively communicate to employees more about business plans and conditions
3. Provide more flexibility in work assignments
4. Delegate more discretionary decision making to lower organizational levels where the information and expertise exist
5. Involve employees in more activities that affect them and require their support and cooperation

Teamwork, communication, and involvement form the human resource foundation for successful automation installations. They must be diligently applied and constantly improved to build and maintain a strong FMS team.

THE SUPERVISOR'S ROLE

The supervisor's role in a conventional batch-oriented manufacturing plant covers a wide variety of activities, functions, and responsibilities. Generally, this involves being many things to many people over and above "supervising" day to day shop floor activities. First-line supervisors or shop foremen often end up being planners, schedulers, coordinators, expeditors, inspectors, tool engineers, part programmers, parts movers, laborers, and whatever other functions are required to support production. In addition, first-line supervision must interface with a variety of engineering, manufacturing engineering, production control, and other support personnel, while receiving rapidly changing and sometimes conflicting demands from higher management. Supervisors must be familiar with all machines in their departments, including all related processes, setups, detail parts, tools, fixtures, item locations, people, and procedures. While concerning themselves with meeting increasing and constantly changing day to day job demands and keeping machine operators busy, the first-line supervisor is also supposed to find time to be disciplinarian, career counselor, helper, administrator, and trainer.

In such an environment, preventive maintenance and teamwork are virtually unheard of as the supervisor is concerned with each individual machine operator's workload, schedule, and production rate, and crisis management occurs with each machine breakdown.

With today's increased emphasis on FMS and factory automation in general, the role, as well as the description of the supervisor, is changing. Descriptively, the typical bull-of-the-woods type is fast becoming extinct and being replaced more and more with well-educated and people-oriented team players.

With increased emphasis on automation, the supervisor has a greater role in managing the interface between the work team and support personnel. The biggest change, however, in the supervisor's role, brought about by automation, is one from giving direction and assigning work to being a *facilitator* and devoting more time to planning, training, monitoring, and helping. This new role (Figure 5-7) is precipitated by the *team* approach. When operators and support personnel are organized into teams with a greater degree of self-management, supervisors become less concerned with internal team regulation and more involved with assisting the team attainment of their production and quality goals. In some factories, self-managing teams have even eliminated the position of first-line supervisor.

With the advent of advanced manufacturing technologies such as FMS, the supervisor is now responsible for a greater concentration of expensive equipment, invested capital, and critical processes. Such increases in responsibility require more significantly increased supervisor education, skills, and abilities over yesterday's bull-of-the-woods type. With FMS and other broad-based factory automation programs, computer-based information systems now tell first-line supervisors the status of all manufacturing processes on a real-time basis. Such finger-tip available information (Figure 5-8) now provides first-line supervision with the capability to make real-time scheduling and priority decisions, while closely monitoring quality, production, and equipment status. Access to such real-time information now requires quick responses to such scheduling changes or unpredictable

Figure 5-7   **The supervisor's role is changing from giving instructions and assignments to being a trainer, facilitator, planner, and helper. (Courtesy of Kearney and Trecker Corp.)**

disturbances to production flow or its support activities. Additionally, such accessibility to real-time data for decision making raises questions about the need and value of the next higher level of supervision, which becomes noticeably redundant unless augmented with higher or additional responsibilities.

FMS and the teamwork approach also mean first-line supervision is given more multifunctional responsibilities in the areas of increased process control, quality control, and maintenance. In a conventional shop environment, if a machine tool breaks down, maintenance would be called, and a repair technician, when available, would be called to correct the problem. In an FMS environment, waiting for a repair technician to "eventually" resolve the problem is unacceptable: hence, the teamwork and more knowledgeable supervision requirements. The automated environment, because of the large amount of capital investment, forces a shift in emphasis from after the fact maintenance involvement to one of preventive and anticipatory problem detection.

As discussed earlier, supervisors in an automated environment like FMS require a broader range of competencies than their predecessors. In addition to being very knowledgeable about the technology itself, supervisors must have a conceptual and practical grasp of the technical, human, and business aspects of production and inventory control. Additionally, they must apply their talents

Figure 5-8 **Computer-based information systems now tell first-line supervisors the status of all manufacturing processes on a real-time basis. (Courtesy of Kearney and Trecker Corp.)**

continuously to anticipating and bringing about change in the organization. Such increased competency requirements places high value and importance on the supervisor selection process. Extreme care and objectivity must be taken to select managers and supervisors whose values and competencies are consistent with the automation objectives of the plant. Highly sophisticated and technically complex automation programs like FMS have no room for ''good ole boy'' appointments to these important managerial positions.

If the overall plant operating rationale includes teamwork, worker participation, group problem solving, and open communication, selected supervision should be able to function capably and comfortably in an FMS environment.

PERSONNEL SELECTION

The selection of good people is the single most important human resource factor in making any factory automation project successful. Selection shortcuts, cannot be tolerated in this critical process. Failure to qualify and assign good people to an automation project like FMS can result in unfavorable consequences much more quickly than in a conventional manufacturing organization.

Obviously, good people alone cannot guarantee project success. Projects can still be poorly planned, ill conceived, lacking in commitment, or inadequately staffed or funded. However, if the selected FMS team is ill equipped to do the job and lacks true desire, no amount of brilliant planning, excess capital, or additional personnel will save the situation. Elaborate computerized control systems will only highlight the inevitable and accelerating approach of failure.

An FMS or any factory automation project should never be staffed with "who's available" type people. It is again very important to make this point because so often people really are selected for project assignments because they "happen" to be available. Consequently, a project of enormous capital investment and importance is jeopardized because the right people are unavailable or a rigorous selection process was bypassed.

In the past, seniority played a major role in advancement and new assignments. FMS and other automation technologies demand a different approach to the selection process. Seniority and experience are becoming less important than the ability to learn. Also, a lot must be learned that was never before part of the worker's job, such as rotating assignments and problem solving in teams. Additionally, FMS training starts earlier, it is highly intensive, more expensive, and time consuming, and resources are wasted if the job assignment proves to be too much for the worker. In many cases it is too risky to weigh seniority too heavily in selecting workers for FMS and other advanced technology assignments.

Samples of possible selection processes include:

1. Job descriptions are posted and applicants submit their qualifications based on job requirements. Project team members interview and select individuals by vote based on a predetermined skill inventory.

2. Interested applicants are interviewed by a combination of management and project team members for attitude first. If applicants appear to possess team player and good interpersonal skills, they advance to the next series of interviews to assess technical qualifications.

3. Applicants complete an 8-hour assessment of their technical and interpersonal skills conducted by a local community college. They then complete a lengthy skill-level inventory.

4. Applicants complete similar skill-level inventories and attitude checks but are then placed in simulated problem-solving situations where each is evaluated by trained assessors on how well they respond.

These are only four briefly explained examples of team selection processes. Obviously, many adaptations, variations, or additions could be used equally as well. However, it is important to note that, in increasing instances, seniority is used only as a factor in the event of a tie score on other rated factors or dimensions.

Management and the project team can devise other innovative ways to give weight to criteria other than seniority for selection and gain the confidence of the work force in the fairness of the selection process. If a company gives inadequate attention to its FMS team selection process, it will end up using only seniority by default.

Generally, the FMS team must be staffed with bright, creative people. Investment in an FMS presents new problems that must be solved in new ways by creative, objective, intellectual effort, not reliance on the same old techniques. Good decisions must be made rapidly, and this requires clear, fast thinking. The FMS team must consist of highly responsive, results-oriented people.

Regardless of the particular position to be filled on the FMS team (technician, operator, programmer, system manager, and so on), certain identifiable traits or attributes stand out as desirable. These attributes represent important factors, in addition to others discussed and not discussed in this chapter, to assist in evaluating applicants for FMS team selection. They should not be interpreted as the only factors to evaluate candidates against, but a careful qualitative and quantitative approach using these factors should yield favorable results. Factors to be considered include:

1. *Energy:* does the individual appear to have a high energy level or strong ability to achieve a high activity level?

2. *Judgement:* ability to develop alternative solutions to problems is critical; also the inherent ability to evaluate courses of action and reach logical decisions

3. *Decisiveness:* readiness to make decisions, render judgments, and take initiating action

4. *Flexibility:* ability to modify behavior or emotions to reach a goal; adaptable to changing conditions and assignments

5. *Sense of urgency:* responsiveness to meet deadlines and commitments and resolve problems without hesitation

6. *Planning and organization:* ability to set and prioritize actions required to accomplish goals

7. *Job knowledge:* job-related information learned and retained based on experience relative to assigned tasks and responsibilities

8. *Independence:* taking individual action when necessary based on personal convictions rather than group consensus

9. *Tenacity:* ability to stay with a problem or line of thought through resolution; perseverance

10. *Proficiency:* measurement of how well previously assigned tasks have been completed; thoroughness; competence

11. *Listening skill:* ability to extract pertinent information in oral communications

12. *Risk taking:* ability to take calculated risks based on knowledge, experience, judgment, and confidence level

13. *Team player:* ability to seek group involvement when problems occur and participate toward common goal

14. *Stress tolerance:* stability of performance under pressure and opposition

15. *Sensitivity:* skill in perceiving and reacting sensitively to the needs of others

16. *Initiative:* self-directional and self-starting, takes action beyond what is required

17. *Creativity:* ability to generate imaginative solutions on a situational basis

18. *Work standards:* strong desire to do a good job for its own sake

19. *Problem analysis:* skill in identifying problems, obtaining relevant information, and logically identifying problem causes

20. *Motivation for work:* importance of work in achieving personal satisfaction, and the desire to be results oriented and a productive contributor

These twenty nonprioritized attributes, although generic in nature, can be used as a beginning framework on which to build an FMS team selection criteria. Personnel selection for an FMS or any factory automation team must be given close, careful attention. Attention paid to selecting the right FMS team players can yield great dividends toward overall FMS success.

JOB CLASSIFICATIONS

Ultimately, in any discussion surrounding FMS, factory automation, and human resources, issues are brought up concerning how employee's jobs are designed and classified. FMS and other automation applications provide companies an opportunity to reexamine job design, classifications, and descriptions within their plants and update them to be consistent with the new technologies. In the mid-1970s, many companies started this job reexamination whether or not they were introducing new technology because traditional job designs failed to motivate employees to do their best. Many changes, at that time, such as work teams and multiskilling of workers, led to increased productivity and employee motivation in plants that introduced little or no new technology. Managers of automated systems cannot afford to ignore any innovation that can motivate employees to keep expensive equipment running at peak efficiency.

Although automation in general has displaced some workers and taken away some discrete responsibilities, it has replaced those responsibilities with others on a higher level and covering a broader range of duties. New technologies like FMS require acute alertness, attention to detail, and problem solving, and job descriptions should reflect the new requirements and multifunctional activities.

Generally, advancing technology, broader responsibilities, and fewer people translate into fewer job classifications. Reductions from 200 to 30 on a plantwide basis are not uncommon. With more people in the same classification capable of performing the same tasks, management has greater flexibility in assigning people on an as needed basis within the factory. Such a simplified job classification system also greatly simplifies the administration of a competitive wage system.

Reducing the number of job classifications almost always is accompanied by broadening the scope of work performed by each individual or by increasing the number of machines or the overall territory for which each individual is responsible. Work scope for each worker is broadened to reduce overhead costs by including some functions previously performed by support personnel or by combining jobs to improve overall responsiveness and minimize waiting time for services.

Jobs can be designed and classified so that both management and labor are happy. Workers rotate assignments and increase skills, thereby resulting in more meaningful work and generally higher pay. Management gains work-force flexibility and improves utilization of its human resources. The new job titles and job descriptions brought about by FMS and other automated technologies now reflect the broader range of new skills required to keep systems operational with fewer support personnel. Common titles in FMS applications include system manager, system operator, FMS engineer, system technician, cell manager, and cell operator.

Work scope, relative to job classification, can be broadened, for example, by:

1. Creating a system maintenance technician position and eliminating the differentiation between electrical and mechanical technicians. Although not always feasible, such consolidation creates a highly skilled, responsive group who can be rapidly deployed to diagnose and resolve a particular system or individual component problem. Additionally, having *system technicians* as opposed to separate mechanical and electrical technicians avoids passing the problem diagnosis and cause back and forth between responsible groups and individuals.

2. System operators, who basically load and unload parts, tools, and fixtures, can be taught and assigned additional tasks to broaden their responsibilities and upgrade their skills. Such tasks include routine machine maintenance and service (maintaining adequate coolant conditions, oiling and greasing machines, performing housekeeping tasks), along with simple troubleshooting and debugging (stopping short of getting into machine and control unit elements). Additionally, system operators have increased responsibility for quality, in-process inspection, tool deformation and wear, statistical process control, and simple NC programming to manually intervene in the event of a problem occurrence. In some cases, system operators even have short-range decision-making responsibilities for work scheduling and machine utilization levels. Such broad-based responsibilities for a system operator in an FMS environment eliminate the need for individual machine operators, laborers, inspectors, and industrial engineers normally required in a conventional shop environment.

3. Giving each shift system manager overall system responsibility. This covers responsibility for total system uptime, utilization levels, workpiece quality and scheduling, preventive maintenance, ongoing maintenance, inventory control, tool assembly, delivery, consumption, and managing the work force. In addition, system managers, prior foremen with increased technical skill requirements and responsibilities, must know the host computer and FMS software in order to communicate directly with the system. It is important to note that, regardless of how jobs are redesigned or classified, individuals assigned to operate, service, and support the FMS must stay with the FMS. Assigning individuals to service and support an FMS and then assigning them to service other areas and equipment in the plant dilutes and conflicts FMS team priorities and can adversely affect system utilization levels and performance.

In general, most workers have the basic skills to do more in an FMS environment than they have been asked to do in the past in a conventional shop environment. And most workers willingly accept the additional responsibilities and could be trained to do still more. The basic skills many people use to keep their homes and automobiles in peak operating condition can be applied to automation. However, not every worker wants increased responsibility and challenge. Some workers simply may not be motivated to learn the new skills required for broad-scoped and multiskilled assignments without management developing a culture that values and rewards learning, self-development, and problem solving. Management then must dedicate sincere effort to clarify and communicate its values by backing its words and providing opportunities for workers to get the needed training and education. They must also reward those who accept the additional responsibilities with recognition, advancement opportunities, and higher pay. Without such ''win–win'' conditions in an FMS or any other factory automation project, neither managers not workers have an incentive to make the proposed changes succeed. Additionally, such emphasis on achievement and project success points up again the need for a fine-tuned evaluation and selection process for workers.

EMPLOYEE TRAINING

Over the past several years, training for new equipment purchases was generally considered an afterthought and given secondary consideration. Workers were generally trained informally on the job by OJT (on the job training) or companies relied on the school systems to provide the background workers needed to get up the learning curve and function adequately on the job.

As technology advanced and equipment became more sophisticated and complex, vendors trained a few members of the company's engineering staff or the highest-seniority union production workers, who then informally trained the proper workers. Companies performed little evaluation of training effectiveness and had little interest beyond the transferring of basic skills to perform a task.

The advent of numerical control, FMS, and other factory automation projects, because of their advanced technology and technical complexity, has brought the importance of quality training into the limelight. Companies more and more are coming to the realization that old training practices with new technologies are falling woefully short of their mark, and the entire process of training and educating the work force with advanced systems needs to be more thorough, intensive, and well orchestrated.

Training for automated cells and systems imposes new demands on the entire plant, often producing more interaction between various groups and individuals. This occurs because a team effort is really required to make a sophisticated system like an FMS run successfully, and because overlap occurs in the skills of the managers, engineers, and production workers required to support the system.

In spite of the new technology forcing increased emphasis on training, an area of primary concern is basic skills: the ability to read, write, think and, above all, the ability to learn. The ability to function adequately with the basic skills is the forerunner to learning and adapting to changing manufacturing technologies. The increasing emphasis on basic skills reinforces the view that training has become much more critical in a company's personnel planning and selection process. And, in many cases, the overall dollars and organizational resources that have had to be committed to the training process have been grossly underestimated.

Education involves three aspects of an individual: mental, learning the facts, figures, and related information; physical, learning how to operate an NC machine tool, for example; and emotional, developing a positive attitude toward handling required tasks. Regardless of how advanced technical training is performed, selected individuals must want and be capable of being trained. Selection of the right people is the first training prerequisite for advanced automation technology like FMS.

Generally, with the purchase of an FMS, preparation of the master training plan is the responsibility of the system supplier. The process of developing a comprehensive master training plan with the system purchaser begins shortly after system sale is finalized. However, before the final sale has been executed, numerous questions surrounding training-related issues will be discussed during the negotiation process. The process of developing the FMS master training plan can best be described by means of the chronology depicted in Figure 5-9. Once the master plan has been agreed on and completed, it should be approved, signed, and dated by the FMS purchaser and supplier project managers. This plan now becomes the official training document for orchestrating the FMS training. Subsequent training plan changes and/or revisions should always be reflected by publishing revised copies of the new training plan and passing the information through both project managers.

The completion of a usable and effective training plan (Figure 5-10) is the most important element to the execution of successful training. The important beginning element, as mentioned in the chronology, is knowing exactly what computers, CNC machine tools, and related support equipment are part of the FMS and how the system will be staffed and operated. These two items must be determined by the FMS purchaser in order to make master training plan development time productive. To develop such a usable working training document, several meetings between the FMS purchaser and supplier training professionals may be necessary. Training plan development should be a joint effort between the FMS purchaser and supplier training teams, but responsibility for executing the plan and for successful training clearly belongs to the FMS supplier.

Major items to be considered in the development of an effective FMS training plan include:

FMS TRAINING PLAN DEVELOPMENT CHRONOLOGY

1. FMS supplier training coordinator discusses with FMS supplier project manager details of system sold to customer.

2. FMS supplier training coordinator visits FMS purchaser to discuss preliminary training needs, plans, requirements, and staffing considerations.

3. FMS supplier training coordinator returns home and develops rough draft of system training flow chart and master training plan and sends to FMS purchaser training coordinator for review.

4. FMS purchaser training coordinator reviews system training flow chart and master training plan with FMS purchaser project team. Ideas, changes, and modifications are proposed and sent back to FMS supplier training coordinator.

5. Master training plan revisions and modifications go back and forth between FMS purchaser and FMS supplier as master plan progresses toward mutual agreement. FMS supplier training coordinator may need to visit FMS purchaser training coordinator and project team to finalize details of master training plan.

6. Finalized copies of master training plan and training proposal (course and activity outlines, lesson plans, prerequisites, and the like) should be dated and signed by FMS supplier and purchaser project managers. As training dates may change (high probability based on FMS master schedule changes), new training plans with new dates are prepared by FMS supplier training coordinator and sent to FMS purchaser training coordinator. All training plan changes and any correspondence between FMS supplier and purchaser training coordinators should be circulated through or copies sent to the respective project managers.

Figure 5-9   The process of developing the FMS master training plan is the FMS supplier's responsibility although it involves a series of meetings, discussions, and an overall joint effort between the FMS purchaser and supplier.

1.   Technical Training

Exactly what training classes and hands-on activities are required to make the new FMS users successful? This is determined from the FMS equipment supplied along with the staffing requirements and considerations.

2.   Location

Some training can best be conducted in the supplier's and associated vendor's plants, while other training cannot be performed anyplace else other than the FMS site and when the system is in full operation.

3.   Duration

How long should each class, seminar, or hands-on activity take? What is the maximum and minimum duration of time for each?

4. Schedule

How will training plan implementation and timing correlate with equipment delivery and installation? This is extremely important as training plan dates must coincide with the equipment delivery, installation, and system acceptance schedule.

5. Prerequisites

Individuals who need to be FMS trained may need to start early to obtain required skills and competencies. This may be done in-house or through local community colleges.

6. Structure

Are course outlines and lesson plans available for each of the classes and activities? Do they contain attainable objectives and will participants be evaluated in each area?

7. Who Takes What?

It may not be necessary for every person supporting an FMS to be cross-trained to know and do everything. Knowing who takes what FMS classes and activities is a function of the FMS purchaser knowing how the system will be operated, organized, and staffed. Deciding who on the FMS team participates in what activities can best be accommodated through preparation of a system training flow chart, as seen in Figure 5-11. Such a chart forces the discriminating thinking required to prepare individual class and activity rosters.

8. How Many?

Sometimes the number of participants in each class or activity becomes an issue. Classes, seminars, and hands-on training should be kept as small as possible in order to maximize individual attention, learning, and participation. More classes with fewer participants may need to be offered in some cases to promote more individualized attention and sometimes on second and third shifts. Offering classes on second and third shifts during the installation period helps to avoid installation and training first-shift interference problems and provides better utilization of time and resources over a 24-hour period.

It is important to remember that there should be training before, during, and after system installation in order to maximize learning effectiveness and retention. Also, quality training, not just training, is the real issue. The FMS purchaser's key training responsibility is selecting the right people and making them available to obtain the necessary information and skills. The FMS supplier's key training responsibility is the development, coordination, and execution of the master training plan.

Additionally, the most successful automated cells and systems are those where the purchaser and supplier work together in a true partnership in which mutual trust, respect, consideration, and concern prevail. Effective training is no exception.

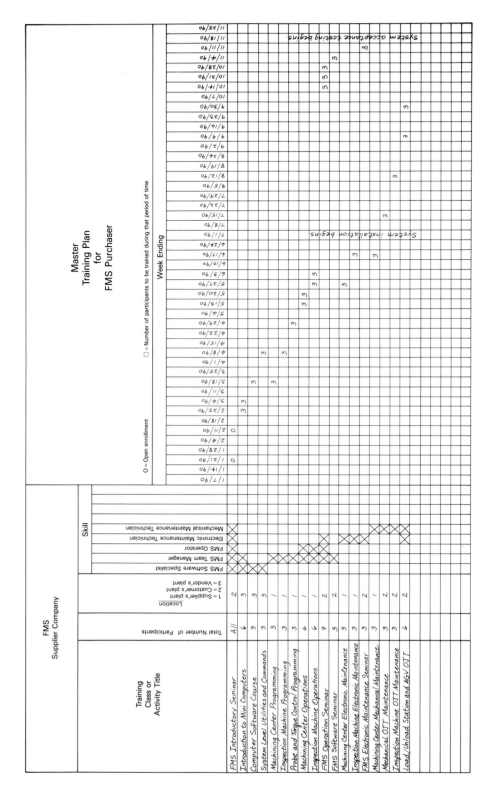

Figure 5-10 The completion of an effective and usable master training plan is the most important element in the execution of successful training.

System Training Flow Chart

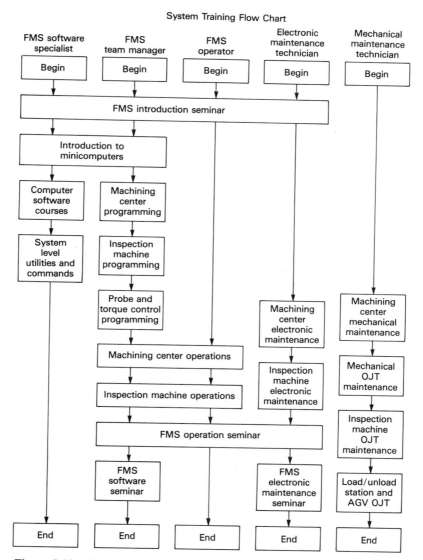

Figure 5-11   Development of a system training flow chart shows who on the FMS team will participate in what classes and activities and in what general order.

SUMMARY

1. Organizing and staffing for automation require changes in the way a company organizes its plant and manages its people.

2. With FMS, people's skills will broaden, but fewer jobs in the FMS arena will be required.

3. Individuals must be more flexible, adaptable, and knowledgeable, performing tasks previously done by several trades and individuals.

4. How an FMS should be staffed depends on how it will be organized and what it will do.

5. FMS requires more technically astute people because of its increased complexity and computerized dependencies.

6. The most successful FMS systems are organized around the team concept of operation.

7. It is much easier to implement new technology when management has established an open relationship with all employees.

8. Management meetings must contain enough useful information about FMS project and implementation plans to be meaningful.

9. The amount of communication effort required to convey project information is directly proportional to the size, scope, and complexity of the project itself.

10. The principal benefits of an FMS team approach are employee multiskilling, flexibility, a leaner organization, and increased job challenge.

11. The central idea of the team approach is to provide relatively few rules and regulations, leaving team members as much discretion as possible in pursuing the teams preestablished goals.

12. Teamwork, communication, and involvement form the human resource foundation for successful automation installations.

13. The biggest change in the supervisor's role is from giving direction and work assignments to being a facilitator, planner, trainer, and helper.

14. The supervisor in an FMS environment is responsible for a greater concentration of invested capital and critical processes.

15. FMS supervisors must be multifunctional, understanding a broader range of technical, human, and business aspects.

16. The selection of qualified people is the single most important human resource factor in making any factory automation project successful.

17. A critical factors criteria list should be assembled and used as an aid to ojectively selecting qualified FMS personnel.

18. Advancing technology, broader responsibilities, and fewer people translate into fewer job classifications.

19. Reducing the number of job classifications almost always is accompanied by broadening the scope of work performed by each individual.

20. In general, most workers have the basic skills to do more in an FMS environment than they have been asked to do in the past in a conventional shop environment.

21. The entire process of training and educating the work force with advanced systems needs to be more thorough, intensive, and well orchestrated.

22. Training for FMS and other automation projects imposes new demands on the entire plant, often producing more interaction between various groups and individuals.

23. Training and education involve three aspects of the individual: mental, physical, and emotional.

24. The completion of a usable and effective training plan is the most important element in the execution of successful training.

25. When actual training occurs (before, during, or after FMS installation) is important to maximize learning effectiveness and retention.

26. The FMS purchaser's key training responsibility is selecting the right people; the FMS supplier's key training responsibility is the development, coordination, and execution of the master training plan.

REVIEW QUESTIONS

1. In an FMS environment, individuals and individual responsibilities are reduced. True or False?

2. _____ , communication, and involvement form the human resource foundation for successful automation installations.

3. How an FMS should be staffed depends on:
   (a) How many machine tools it will contain
   (b) How many people will be displaced as a result of implementing FMS
   (c) How it will be organized and what it will do
   (d) The system's complexity

4. FMS workers, in general, must be more flexible and responsive and have broader-based skills than their conventional machine shop counterparts. True or False?

5. An FMS system operator requires an in-depth knowledge of host computer and FMS software commands, file structures, and functions. True or False?

6. Employee meetings are successful and communication has occurred as long as meetings occur on a regular basis. True or False?

7. Getting people involved through a participative management style is easier with an existing plant than with a new plant start-up. True or False?

8. By organizing FMS responsibilities around teams:
   (a) Ongoing maintenance and machine uptime become everyone's responsibility
   (b) Responsibility is increased for quality
   (c) The focus of problem anticipation and prevention is a team-measured goal
   (d) Only (a) and (c)
   (e) All of the above

9. Discuss what changes are required in the supervisor's role to adequately function in an FMS environment.

10. Selection of good people is the single most important human resource factor in making any factory automation project successful. True or False?

11. List ten important personal attributes or characteristics to check for during the FMS employee selection process.

12. Reducing the number of job _____ almost always is accompanied by broadening the scope of work performed by each individual.

13. Management's gain by reducing job classifications and broadening employee's skills is increased work-force flexibility and better utilization of human resources. True or False?

14. Training, with automation projects like FMS, is of primary importance because:
    (a) Individuals will not take the initiative to learn the necessary skills unless required to do so
    (b) A team effort of multifunctional people is really required to make an FMS perform successfully
    (c) People like to know what's happening in the field of automation technologies

15. Education involves three aspects of an individual: mental, physical, and _____ .

16. Choosing which FMS training classes and activities employees should participate in is the first training prerequisite for advanced automation technology like FMS. True or False?

17. Preparation of the _____ is the most important element in the execution of successful training and is the responsibility of the _____ .

18. Where FMS training classes are to be conducted (location) is not an important consideration to overall learning and training effectiveness. True or False?

19. An FMS software seminar would typically be conducted (a) before, (b) during, or (c) after FMS installation.

20. The FMS purchaser's key training responsibility is:
    (a) Evaluating training delivery and effectiveness
    (b) Selecting and making the right people available for training
    (c) Preparing the master training flow chart
    (d) Deciding which classes should be taken and how many people should participate

Quality: Manufacturing's Driving Force

6

OBJECTIVES

After studying this unit, you will be able to:

☐ Describe quality and its relationship to upfront planning.

☐ Identify five characteristics of quality.

☐ Name and briefly explain two methods for getting and keeping quality under control.

☐ Describe the importance and relationship of quality to flexible cells and systems.

INTRODUCTION

Quality is the most important factor to the survival of any business. It directly supports the other factors of cost, productivity, on-time delivery, and market share. In a free enterprise system, quality standards are set by the customer. Therefore, all quality standards needed to produce the components of a product and perform its assembly must be specified in a manner such that customer expectations are met.

The goal in any business then is to improve the quality of quality. This means establishing quality of product and quality of process standards from product and design engineering through manufacturing and assembly to customer acceptance and serviceability. It also means applying attention to detail and using common sense.

Fixing manufacturing problems is key to improving quality and meeting customer expectations, even though the process starts further upstream. Manufacturers are coming to realize that they must have better control over difficult to control manufacturing processes. Flexible cells and systems are a means of improving and controlling the manufacturing process, but only after the process itself is under control and all the required upfront planning and preparation work is completed.

Pressure in the manufacturing industries continues to mount to "get it right the first time"—to design products correctly and to make them of higher quality, more quickly, and less expensively than ever before. Flexible cells and systems afford users an opportunity to accomplish these goals, not as an end in themselves, but as a means to meeting the challenges of a cost, quality, and delivery end.

DEFINITION, DESCRIPTION, AND CHARACTERISTICS

Throughout industry, quality is being viewed as the most powerful corporate leverage factor for achieving both customer satisfaction and lower costs. This renewed emphasis on quality is occurring at a time of increased global competitiveness and an understanding in all types of business and industry that, in order to survive, quality must be number one.

Nowhere is the reemphasis on quality greater than in the manufacturing industries where raw material is converted to finished product. No longer is the focus only on the product or process; it is on the customer. No longer is the responsibility only on the manufacturing and quality control people; it belongs to marketing, design engineering, manufacturing engineering, and perhaps, most importantly, top management. And no longer does quality only mean solving problems; it also means preventing them and creating opportunities.

Quality in the past meant traditional "after the fact" inspection or testing that was designed to catch defects only after they occur. Such quality-control and inspection programs have been implemented on a defect-driven basis, not on a prevention basis as encouraged by the quality experts like J. M. Juran, and W. Edwards Deming. Quality, on a prevention basis, means doing it right the first time and taking the upfront planning steps to ensure product and process quality.

Quality is the least understood but most important factor in manufacturing. It is the least understood because many people have misconceptions about what quality is and what it is not (Figure 6-1). Understanding the differences can be critical to long-term business success or failure.

Simply mandating defect-free parts, zero-defect programs, and quality

WHAT QUALITY IS	WHAT QUALITY IS NOT
Customer satisfaction and fulfilled expectations	Finding bad parts
Everyone's Responsibility	Something that the top tells the middle to do to the bottom
Giving people the tools and training they need to do their jobs	Signs, posters, and lip service about the importance of quality
Keeping the process under control	Inspecting quality into the parts
Controlling quality at the source	Adding more inspectors
Defect prevention driven	Defect inspection driven
Meeting the demands of the customer	Meeting the demands of the inspector
Doing it right the first time	There's never enough time to do it right but always enough time to do it over
Continuous improvement	Relaxing improvement efforts after achieving defect reduction goals
Solidly supported by an involved top management	Delegated to first-line supervision
A concept integrated throughout the organization and in all company operations	Demanding quick fixes and defect-free parts
Supported by quality control tools	Driven by quality-control tools
80% management controllable	80% worker controllable

Figure 6-1 Many misconceptions exist about what quality is and what it is not

awareness programs by top management will not get the job done. Improving quality is not something that the top tells the middle to do to the bottom. In fact, close analysis of manufacturing mistakes in most organizations will reveal that the cause of 80 to 85 percent of quality-related problems is not in the individual worker's control. Examples include:

□ Lack of part design for manufacturability and assembly
□ Purchasing and procurement mistakes
□ Suppliers sent defective parts
□ Lack of adequate worker training
□ Lack of proper tooling to do the job
□ Obsolete or worn-out machinery and equipment
□ Parts damaged in handling
□ Faulty production processes
□ An "I need the part, we'll make it fit at assembly" attitude

Such findings vividly support the Juran and Deming statement that "eighty per cent of quality problems are management controllable."

What is quality? Although responses to this question abound, the answer that cuts to the core is: *Quality is meeting the requirements and expectations of the customer.* Generally, defining customer requirements is not that difficult. The customer simply wants:

□ Highest quality
□ Lowest price
□ Fastest delivery
□ Greatest flexibility to meet special concerns and considerations
□ Least maintenance with limited or no service

Quality is and should be defined by the customer. That is, the customer wants products and services that, throughout their life, continue to meet functional needs and expectations at a cost that represents value. This is a fundamental change in the way quality used to be considered. In the past, quality was often determined on the basis of what marketing or sales thought was the customer's need, along with "this is what I've designed" from product engineering. Now, however, quality is based on buyer preferences, continuous cost improvements, and value, all of which means manufacturers are under more pressure to stay competitive and profitable, yet fulfill increasingly demanding customer requirements.

However, quality for the most part is still measurement, but now it's a question of where those measurements are taken, before or after the process. By regularly reviewing actual performance against plan in critical work areas such as inventory, production scheduling, product delivery, and customer service, management can pinpoint quality problems and problem areas before they adversely affect bottom-line results. Reducing errors and maintaining product uniformity and efficient production are still vital, but quality is now viewed as an attitude of improvement that must contagiously penetrate the entire organization.

Turning attention to meeting customer needs and expectations requires clarification of the word "customer." Most people think of customers as the individuals who buy the end product or service. But, actually, even though the customer in the purchasing marketplace is the external end target, there are many internal customers that first must be satisfied in order to ultimately satisfy the purchasing external customer. For example, assembly is manufacturing's customer, manufacturing is engineering and design engineering's customer, and so forth. These internal customers are employees who must rely on the workmanship, craftsmanship, and leadership of other employees in other departments to effectively do their jobs. This is an extremely important concept, as every function in the organization has customer–supplier relations that critically affect the operations of the company and therefore its presence in the marketplace.

As quality concerns heighten, manufacturers are becoming increasingly aware that focused attention on quality can mean as much or more to the bottom line as do increases in production and quick responses to market changes. And the quality of the end product and how it is perceived by the customer is the sum total and direct result of the internal attention paid to quality.

To improve quality, it is important to focus on the individual characteristics that redefine quality. Although individual companies and their products are as different as individuals, those differences fade and commonality becomes clear when attempting to meet customer needs and expectations. In general, the characteristics of quality are:

1. Reliability

Customers want reliable products. Unreliable products or processes cause customers to lose money and time and nobody likes to be inconvenienced. Costs may extend beyond the product itself for damage done to other equipment and additional loss of time, money, and potential market share.

2. Delivery

On-time delivery is one of the highest customer expectations. Product life cycles continue to shorten, making on-time delivery even more important and difficult to achieve. Astute planning and realistic scheduling are critical to achieving on-time delivery.

3. Manufacturability

Products that are designed to be efficiently and effectively manufactured and built are critical to achieving quality. Easily produced products reduce delays and cost. The key to achieving manufacturability is through integration of engineering and manufacturing.

4. Communication and Involvement

For quality to flourish, it must be anchored in an environment of open and honest communication and active employee involvement. Training is also a very important aspect of quality and a continuous improvement process. The reason is very clear: skilled and informed people create quality products; unskilled and uninformed people create problems.

5. Serviceability

Customers first of all want products that do not need to be repaired. When repair is required, it needs to be quick and simple. This is where design for manufacturing and assembly pays off again. Improved serviceability also comes from quality manuals, documentation, and diagnostic information to promote easier pinpointing of failures and quick fixes.

6. Cost

A very important customer consideration and quality characteristic, cost means more than just price; it means real value and perceived value for the dollar. Arguably, cost is distinct from quality, but in a customer's mind, it is value for the dollar. Cost is also a function of a company's continuous improvement efforts. Customers are demanding and getting yearly value increases and price decreases.

7. Functionality

Synonymous with performance, functionality is the most common measure of quality. Although difficult to improve, product functionality can never be compromised because it can easily be compared with a competitor's product.

8. Consistency

Also a function of continuous improvement, consistency must not only be predictably constant, but it must improve. Loss of consistency and predictability can be death to a product and its manufacturer. Simply maintaining consistency for consistency's sake can be equally destructive; continuous improvement is the only answer.

GETTING QUALITY UNDER CONTROL

Getting quality under control before any automation improvement program is begun is extremely important. Without such a disciplined approach, defective parts are just made more productively. So the first and most important step in getting quality under control is to realize that low-tech improvements and control must be done before high-tech systems are installed.

Control, in the words of quality expert J. M. Juran, means staying on course, adherence to a standard, prevention of change. Although the goal should be continuous improvement, which means continuous change, getting control of existing product and process quality factors is the unavoidable next step.

Many tools are available for determining a corporation's quality status and for seeking improvement. Some of these include scatter diagrams, histograms,

Pareto charts, data collection, and cause and effect analysis. Many of these were and still are used in conjunction with quality circle and productivity–quality improvement programs. However useful these tools are and continue to be, one tool, statistical process control, or SPC for short, continues to play a major role in achieving and maintaining quality.

SPC is not a new quality-control tool or gimmick; it has been around since the late 1920s, but has finally come into its own as a result of unprecedented global competition. Its basic principle lies in the fact that when a large number of any items are measured no two of them will be exactly alike. However, all will fall within a standard and predictable deviation from an average. This distribution predictability will occur regardless of the number of characteristics that can be measured.

As measurements are taken on any item, most of the measured characteristics will be clustered around the average, with some measurements on the high and low limits of the range. If these measurements are plotted on graph paper, the result will be a bell-shaped curve, called the frequency distribution curve, as illustrated in Figure 6-2. Through statistical analysis of SPC then, a prediction of where 68, 95, and 99 percent of the parts will fall on that curve can be made. This is accomplished by measuring and determining the average (x-bar) and the standard deviation away from that dimension, either plus or minus. This deviation is called sigma, and SPC uses three sigmas on either side of the x-bar average for a total of six, as seen in Figure 6-3. This will determine mathematically where 99 percent of the measured parts will fall within the normal distribution curve.

The importance of this concept is that, once SPC is established, inspection need only occur on one or two parts every half-hour or so (depending on produc-

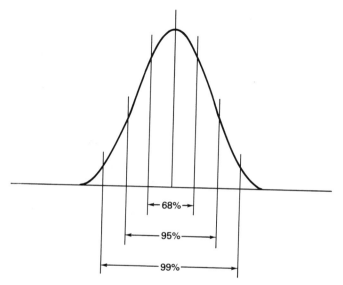

Figure 6-2   Approximately 99 percent of all manufactured parts will fall within this frequency distribution.

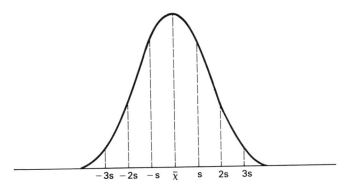

Figure 6-3 The frequency distribution or bell-shaped curve indicates the standard deviation of plus or minus three sigmas.

tion rates) in order to make sure production requirements stay within tolerance. Therefore, it is not necessary to inspect every part. If a part deviation does occur, steps are taken immediately to uncover the cause of the defect and fix the problem instead of waiting and generating scrap or rework. In modern SPC systems, all measuring devices are connected to computers and substitute a CRT screen for graph paper and manual entry.

Although presented here in an overview manner, the power of SPC as a tool for control quality should not be underestimated. It can help prevent scrap and rework instead of detecting it after it has been made, and it is a method of knowing and documenting that parts meet specifications. It continues to gain increasing use and widespread acceptance as a primary tool to get quality under control.

Another applicable tool for getting quality under control that merits discussion is Taguchi methods. Pioneered by Genichi Taguchi of Japan, Taguchi's methods are accepted and used by many corporations throughout the world. A multiple winner of the Deming prize for his writings on quality, Taguchi's methods focus on dollars, not standard deviations, reducing variation and loss between products, and eliminating problems upfront in the design phase. Using the Taguchi *quality-loss* function, losses can be expressed mathematically in dollars. The net result is that manufacturers can get a real picture of where their products and processes can be improved and the cost of improving them, even when the processes conform to specifications. As a result, there is strong motivation to continuously improve.

The most important aspects of Taguchi methods are to eliminate problems at the source by creating designs based on actual customer usage conditions. Careful upfront consideration would be given to customer requirements, actual usage, and variation under suboptimal conditions. Engineering's goal would be to determine the optimal combination of design factors that gives the least variation under anticipated customer conditions and to increase quality as much as possible without increasing cost.

Further study and application of Taguchi methods can play a major role in identifying what factors of the process are critical to producing the levels of quality required in an organization. Additionally, Taguchi methods are a vital key to automation success because they raise the level of cost consciousness about

---

SEVEN STEPS TO QUALITY IMPROVEMENT

1. Make astute planning part of the quality improvement process.

2. Every defect, every problem is an opportunity for improvement.

3. Identify and prioritize the vital few versus the trivial many quality defects and problems.

4. Control quality at the source.

5. Make all workers do their own rework.

6. Increase quality awareness and improve quality skills of all workers through rigorously applied education and training.

7. Promote communication and involvement through cross-functional team building and problem solving.

Figure 6-4   Staying focused on quality improvement can play a major role in getting and keeping quality under control.

automation decisions and automation's impact on quality, and because they are prevention rather than detection oriented.

Regardless of what methods are taken to improve quality, the important thing is that quality does not just happen; it must be planned for, controlled, and well designed before it can be well built. Studies indicate that only about 20 percent of quality defects can be traced to manufacturing; the other 80 percent are locked in during the design phase or by purchasing policies that value low price over the quality of parts and material received. It must also be realized that every defect, every problem is an opportunity for improvement. Defects and problems should be identified, ranked by priority, and a rigorous effort be made toward continuous improvement. And, finally, staying focused on the seven steps to quality improvement as outlined in Figure 6-4 can play a major role in getting and keeping quality under control.

IMPORTANCE TO CELLS AND SYSTEMS

Quality, as we have seen, is a concept that is integrated throughout the organization, beginning at the supplier and raw material stage and design engineering and continuing through manufacturing engineering, manufacturing, distribution, and service. Quality programs that stick become an attitude and an organizational way of life. They cannot be "flash in the pan" programs that lack financial support and solid management commitment and are discontinued when production demands increase or business falters.

Quality is a by-product of caring and, when viewed as everyone's responsibility, should have just as much emphasis, and maybe more, on the upstream functions of design engineering and procurement than on the downstream functions of manufacturing. Quality of design must be measured in terms of reduced cost, better use of materials, reduced complexity of manufacturing, and increased product functionality. However, fixing and improving manufacturing is the key to improving quality. Quality improvements should begin in the manufacturing middle, allowing improvements to spread in both directions at once—upstream into design and product engineering and downstream into support, service, and distribution. Regardless of how and where a product is built, manufacturers are beginning to understand that they must have better control over their manufacturing resources. Better control means more control, and automation and innovation provide both more and better control in the manufacturing arena.

Quality and productivity are inseparable with regard to automation, particularly flexible cells and systems. Generally, the two remain distinct and conflict with one another, as typically described by manufacturing's definition of quality-control people—the ones who keep manufacturing from shipping parts. However, in an automated cell or FMS, the two are bound together as in marriage to achieve the goal of a successful union and relationship.

Quality is especially important to flexible cells and systems because:

1. It is a major benefit derived from a cell or system installation. Many decisions to automate are based on quality improvement as being one of the key justification factors.
2. Cells and systems bring predictability and consistency to the manufacturing process, thereby adding quality improvement through repeatability and reliability.
3. Human intervention and decision making are reduced as a result of automation; therefore, quality is improved once the processes are under control because many of the day to day error-prone human elements have been eliminated.
4. Automation aids quality improvement by virtue of its forced attention to planning and detail. Emphasis on automating and controlling the process focuses attention on the quality of the end product and of the entire process.
5. Automating and improving quality reduces scrap and rework if processes are under control prior to automation.
6. Quality means continuous improvement, and flexible cells and systems are a form of continuous improvement.

Quality is also teamwork, internally in an organization or externally between customers and suppliers through concurrent engineering and open, honest communication. Particularly with cells and systems, quality of teamwork plays a very important role because of the many types of equipment and vendors involved.

Quality of teamwork begins in the planning process for any automation improvement project between system user and system supplier and continues through training, installation, acceptance testing, and implementation. Quality of teamwork is the best kept secret of a successful cell or FMS installation.

SUMMARY

1. Quality in the past was defect detention driven but now is defect prevention driven and means doing it right the first time.

2. Quality is meeting the requirements and expectations of the customer, and it is the least understood but most important factor in manufacturing and business survival.

3. Quality is and should be defined by the customer and is based on buyer preferences, continuous cost improvements, and value.

4. The quality of the end product and how it is perceived by the customer is the sum total and direct result of the internal attention paid to quality.

5. To improve quality, it is important to focus on the individual characteristics that redefine quality.

6. The first and most important step in getting quality under control is to realize that low-tech improvements and control must be done before high-tech systems are installed.

7. Many tools are available for determining a corporation's quality status and for seeking improvement; however, statistical process control (SPC) and Taguchi methods play a major role in helping to achieve and maintain quality.

8. Quality improvements should begin in the manufacturing middle, allowing them to spread upstream into design and product engineering and downstream into support, service, and distribution.

9. Quality and productivity are inseparable and very important to flexible cells and systems.

REVIEW QUESTIONS

1. Quality is:
   (a) Delegating responsibility to first-line supervision
   (b) Eighty percent worker controllable
   (c) Finding bad parts
   (d) Defect prevention driven

2. For the external buying customers to be satisfied, the demands of the internal customers must first be met. True or False?

3. The quality characteristic of _____ is important because customers want products that are designed to be efficiently and effectively manufactured and built.

4. The first step to getting quality under control is:
   (a) Continuous improvement
   (b) Finding and preventing defects
   (c) Instituting quality awareness programs
   (d) Realizing that low-tech improvements must precede high-tech automation installations

5. SPC is a new quality tool designed to predict deviation from an average. True or False?

6. The standard deviation in SPC away from the (x-bar) average is known as _____, and in a normal bell-curved distribution there are a total of _____.

7. The most important aspect of Taguchi methods is that they are focused on eliminating problems at the source. True or False?

8. Discuss three reasons why quality is especially important to flexible cells and systems.

7

Just-In-Time Manufacturing

OBJECTIVES

After studying this unit, you will be able to:

☐ Describe the concept of just-in-time manufacturing and explain its relationship to FMS.

☐ Identify five JIT benefits.

☐ Name the two JIT implementation cornerstones.

☐ List four quality and four quantity JIT principles.

INTRODUCTION

The stockless production concept of just-in-time (JIT) manufacturing, originally pioneered by the Japanese, is about inventory, but it is much more than inventory reduction. It is about organizing the production process so that usable parts, both purchased and manufactured, are available on the shop floor when they are needed—not too late and not too soon.

Although inventory reduction is one of the benefits of JIT, frequently it is not the major benefit. JIT has helped many companies reduce quality defects by as much as 60 percent, decrease production time by as much as 90 percent, and cut capital expenditures by as much as 30 percent when the technique is applied to the full circle of procurement, manufacturing, and delivery.

Work-in-process and finished goods inventories are an expensive luxury, but many companies are doing well and do not have the economic incentive to get on either the JIT or automation bandwagon. However, ignoring the JIT concept is ignoring an opportunity and a chance to make invisible problems visible where they can be squarely faced and resolved.

FMS fits well within the concept of JIT as the goals and concepts of off-line part setup, increased flexibility, and improved quality and part throughput are common to FMS and JIT. Therefore, planning and implementing an FMS can and should be a driver for changing a company's total manufacturing approach to one of just-in-time.

DEFINITION AND DESCRIPTION

Just-in-time (JIT) is the name commonly used to describe a stockless production manufacturing approach where only the right parts are completed at the right time. To have the exact quantity of parts when they are needed and where they are needed enables a firm to control raw material, work in process, and finished goods inventory and to know exactly what will be available for shipment.

JIT is an operation philosophy and an operation strategy, not a system. You cannot go out and buy a JIT system. JIT is not a destination but a journey, achieved by implementing a variety of techniques of continuous improvement (see Figure 7-1).

In today's manufacturing environment, most companies are confronted with fierce competition both domestic and offshore, in terms of quality, delivery, and price. Customers demand quality, expect on-time delivery in exact quantities, and usually ask for yearly price decreases. Consequently, the primary goal is to reduce costs while improving overall quality. Reducing inventory levels is a most effective way of controlling costs and, at the same time, controlling quality and delivery times.

The obsession with inventory reduction originates from an understanding that inventory ties up money the corporation could better use elsewhere and knowledge that high inventory levels are used in many cases to cover up other problems. JIT manufacturing is about organizing the production process so that inventory of raw materials, work in process, and finished goods are driven to a minimum and kept under control. The primary goal is to have only the lowest necessary level of material arrive at the plant (or be made within the plant) exactly when it is ready to be used.

Many aspects of production need to work together effectively if this goal is to be reached. Master production scheduling, capacity and material planning, purchasing, shop floor layout, and manufacturing engineering are some factors that must change if a JIT approach is to succeed. Additionally, JIT must be viewed as an opportunity for change and not a threat to change. Creating such an environment for making the operating changes necessary to implement a JIT manufacturing approach can be a bigger challenge than the actual changes themselves.

JIT IS:

□ A stockless production manufacturing approach

□ An operation philosophy and an operation strategy

□ Not a destination but a journey

□ Reducing inventory, improving quality, and controlling costs

□ A concept where the primary goal is to have only the lowest necessary level of material arrive at the plant or to be made within the plant exactly when it is ready to be used.

□ An opportunity for change and not a threat to change

□ A "pull" system where parts are produced only when they are required

□ Sometimes viewed as a manufacturing revolution

Figure 7-1 Just-in-time manufacturing is a new approach to organizing the production process.

In a typical batch-oriented factory, parts are bought and produced in mass quantities, taking advantage of quantity discounts and already made setups. This strategy on the surface appears efficient as parts are being ''pushed'' through the system due to the economies of scale of purchasing and manufacturing. A ''push'' system feeds on itself by forcing parts to be completed faster, only to wait longer for use. Parts are usually wanted early due to insecurity caused by fear of not getting the parts on time. These factors lead to greater inventories and an overreliance on the informal production and inventory control system and hot, hotter, and hottest levels of part expediting.

JIT, on the other hand, is a ''pull'' system where parts are produced only when they are required, even if it means idling machines and workers for a period of time. Components are bought and parts are produced and ''pulled'' to the next work station based on need. No parts are produced until an authorization, often referred to as a *kanban*, is issued from the next work station. A kanban is a Japanese word that implies open authorization. This means that, since the next work station is open and a part is required, work can begin at the present work station. If the next work station is blocked because parts are waiting to be processed, work at the present location waits for the kanban authorization from the next work station.

JIT in many instances is being viewed as a manufacturing revolution that holds the key to strengthening a company's competitive position, freeing up working capital, and improving product quality, while reducing cost and lead time. It is being compared to the 1908 revolution when Henry Ford introduced the assembly line concept that dramatically changed how automobiles and other products are built.

BENEFITS AND RELATIONSHIP TO FMS

Although the concept of just-in-time manufacturing is clear and understandable, many manufacturers do not understand the full benefits of JIT and are reluctant to implement this approach. This is because:

1. The emphasis of JIT literature is directed at inventory management, not the total effects of JIT on the overall business.
2. Business is good for many manufacturers, and many may not yet feel the effects of offshore competition and therefore see no short-term threat.
3. Most companies react to threats, and if there is no threat to a company's existence on the competitive basis of cost, quality, or delivery, then management has little incentive to act.

The facts are that JIT benefits come from a variety of sources, but the majority of companies are just looking at material flow improvement, which is only a small part of what JIT can bring to a company. It can shorten product-development times, improve order entry cycle times, and improve each step of the production cycle. In addition, JIT can decrease the amount of time a company holds raw materials and finished products, and it can reduce the time from cutting an invoice to collecting cash for sold products.

Implementing a just-in-time manufacturing plan can achieve a wide range of benefits and results with relative ranges consisting of:

- □ 70 percent reduction in departmental part moves
- □ 60 percent reduction in lead time
- □ 60 percent increase in overall throughput
- □ 60 percent reduction in tooling repair cost
- □ 55 percent reduction in complete job changes
- □ 45 percent reduction in material handling
- □ 40 percent reduction in machine downtime
- □ 40 percent reduction in total inventories
- □ 40 percent decrease in job classifications
- □ 40 percent reduction in setup times
- □ 30 percent reduction in rework
- □ 25 percent increased capacity
- □ 20 percent reduction in floor space
- □ 10 percent reduction in direct labor

Such changes in manufacturing have resulting effects on assembly operations, consisting of:

1. Supplying assembly with the required parts as needed; requirements are "pulled" through production according to assembly demand
2. Building products in relationship to market demand and received orders, not for stock
3. Reducing the number of assembly line "quick-fix" changes and conversions

Although one of the principal benefits of JIT is, in fact, inventory reduction, inventory does perform a valuable economic function, at times. Acceptable stock inventory would include:

1. Buffer stock to compensate for variations in vendor deliveries
2. Hedge, or prebuying, to compensate for future price increases or shortages, if the carrying cost to justify the prepurchase is adequate

The benefits of JIT are real and broad based and affect many functional areas within an organization, as seen in Figure 7-2. And these benefits are capable of being achieved, with or without a cell or FMS.

Throughout the text, emphasis is on material flow and part throughput internal and external to the cell or FMS. With FMS, sophisticated software handles the part scheduling for each work station, controls part movement from station to station, handles NC program download to each CNC unit, and performs a variety of other functions. Ideally, well-implemented and operational JIT techniques should be a prerequisite to FMS and be in place before a flexible cell or system is installed in order to follow the "simplify before you automate" rule. Unfortunately, this is not the case in most companies, as adding cells and systems usually precedes major material flow and inventory reductions. JIT can be either a cause or an effect of FMS. However, it is difficult to implement FMS without JIT and achieve the true results of FMS.

FMS forces operational and organizational change in a company. FMS is also a means to a just-in-time end and can be used in many cases to drive productivity improvement changes like JIT and group technology through the organization. The training, teamwork, cooperation, planning effort, and positive attitudes used to implement a cell or system can be carried over and broadened to undertake implementation of JIT techniques. Installing a cell or system first can provide a seedbed for planting and leveraging a JIT discipline.

The installation of a cell or system provides the manufacturer with the flexibility to produce parts in lot sizes as small as one. With an FMS, it is no longer necessary to carry excessively large inventories or issue high economic order quantities in an attempt to satisfy anticipated customer demands. The accuracy of marketing's forecast would become less critical since the manufacturer would now have the option of producing to order. Consequently, the just-in-time philosophy advocated by a flexible manufacturing system would result in decreased lead times,

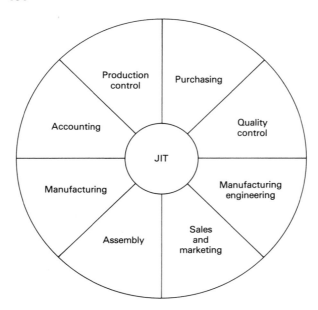

Figure 7-2 Functional areas benefited by JIT.

less work in process on the shop floor, smaller finished parts inventories, and increased customer satisfaction.

IMPLEMENTATION CORNERSTONES

In a detailed discussion of JIT, the two items that emerge as vital to just-in-time implementation success are improved material flow and setup reduction.

Overall material flow is being viewed by more manufacturing personnel as an integral part of the entire automation effort. Material flow means inventory movement, and inventory includes all categories of materials carried in the company's financial statements, including raw materials, purchased components, tooling, work-in-process parts, and finished goods. In addition to these categories, supplies and special nonvalued or used tooling should be included if significant because they must also be moved and are part of the flow process. Overall part flow is dictated by the slowest producing work station or element.

The ideal material flow would be that achieved if a single part were being made. There would be no stopping of that part as it goes through the various sequential production operations. At each work station, the machine would be waiting, the part would be entered into each station, and at the end of each operation, the material transfer device would be ready to whisk it to the next machining station.

However, overall material flow must be improved, and before just-in-time techniques are applied. This involves addressing wait time, and not just trying to ''speed up'' material flow between work stations in order for parts to ''hurry up and wait'' at the next machining station. By improving overall material throughput (both internal and external to the cell or FMS), the period of time that material is owned is reduced, and therefore the investment in inventory.

The goal is to have stock flow uninterrupted from receipt directly through each production step. It follows then that whatever causes material to stop, other than the actual process, is a roadblock target for investigation. Primary roadblock targets of work-in-process inventory investigation should include:

1. Finished parts waiting for stocking operations
2. In-process parts waiting for movement to the next work station
3. Parts waiting to be machined at their current work station
4. Parts waiting for setup components, tooling, or planning instructions

Causes of material flow delay and wait time include:

1. Parts waiting in an assembly queue for a delayed component
2. Ordering before material is required
3. Planned buffer stock to compensate for variations in on-hand balance accuracy, customer requirements, or intermediate production completion time
4. Stock ordered in error due to errors in the on-hand balance or a wrong part number
5. Lack of disciplined engineering change control
6. Inaccurate demand forecast
7. Economic order quantity set in excess of current requirements
8. Batching of workpiece runs to combine setups
9. Late release of shop floor orders, which compounds scheduling problems and adds delay time

The other item to be examined for improvement prior to JIT implementation is setup reduction. Manufacturing plants got into the bad habit of producing workpieces in large batches to avoid frequent and time-consuming setups and teardowns of machines. It seemed to make sense that, once a machine was prepared to produce a certain workpiece, a shop should produce as many pieces as possible. This approach maximized machine utilization and chip-making time and, in terms of parts produced per machine unit of time, was highly efficient.

However, the result was excess work-in-process and finished stock inventories and increased overhead for workpiece handling and storage. Worse than this, such narrow thinking caused inflexibility and unresponsiveness to changing manufacturing conditions and requirements. Engineering changes, in many cases, continued to be resisted, even when product quality was at stake, to avoid obsoleting existing stock. Workpiece changes and defects that were not detected until the part was either downstream or already assembled meant that an entire batch or several batches of workpieces had to be scrapped. At that point, long lead times due to setup seriously delayed recovery.

Figure 7-3   The use of quick-change clamps can aid considerably in reducing setup time. (Courtesy of Kearney and Trecker Corp.)

Fortunately, attitudes are changing and manufacturers are recognizing the importance of setup reduction as an applied technique of JIT. Setup reduction efforts should include:

1. Producing only the minimum necessary units, in the smallest possible production-run quantities, at the latest possible time, with the objective of achieving plus or minus zero performance to schedule

2. Emphasizing simplicity in setups by eliminating as many nuts, bolts, clamps, and fasteners as possible through the use of quick-change clamps, fasteners, pushers, and other available accessories (Figure 7-3)

3. Moving on-line machine setups to off-line palletized preparation

4. Using fixed gages and templates for machine adjustments and to minimize machine "tweeking" to obtain accuracies

5. Standardizing tools, fixtures, and related components as much as possible through the use of modular fixturing (Figure 7-4)

6. Reducing manual efforts such as walking and searching to find needed tools and components

7. Reducing multiple job classifications and making workers multifunctional, increasing their value to themselves and to the company

Figure 7-4 Standardizing tools, fixtures, and related components is another method of helping to reduce setup time. (Courtesy of Kearney and Trecker Corp.)

8. Designing parts for manufacturability
9. Taking advantage of operator know-how through cross-training and involvement and by treating every employee as a "mini-consultant," since the employee knows more about his or her machine and operation than anyone else

Capitalizing on improving plant-wide material flow and reducing part setups are the most important steps to beginning JIT implementation. Undertaking these efforts early in any automation effort is following the cardinal rule of "simplify before you automate!"

QUALITY AND QUANTITY APPLICATION PRINCIPLES

Much has been discussed in this chapter relative to just-in-time manufacturing: what it is, what it can do for a manufacturing organization, its relationship to FMS, and some beginning implementation techniques. However, for JIT to be successful and function properly, part quality and quantity must be under control. This means that:

1. All parts that arrive for assembly must be usable parts.
2. All parts must arrive where they are needed, when they are needed, and in the exact quantities needed.

Problems related to quality in an organization should be addressed before quantity issues because it becomes very difficult to focus on quantity and delivery issues if part quality is questionable. One reason the Japanese have been so successful in the manufacturing world is that they understand this concept very well and have made great strides in solving quality problems before addressing quantity issues.

The primary quality JIT principles are:

1. Train every person in the organization on quality issues and heighten their sense of quality awareness.
2. Make continuous improvement toward zero defects a way of life and make defects visible.
3. Control quality at the source by making each production worker responsible for quality and eliminate trying to "inspect quality into the parts."
4. Make every worker responsible for his or her own rework.
5. Establish preventive maintenance and make production equipment maintenance the responsibility of the production worker using that machine.
6. Encourage teamwork and worker involvement in decision making.
7. Reduce multiple job classifications and make workers multifunctional through cross-training.
8. Qualify vendors, with source inspection and validation required before delivery.
9. Simplify product design so that it incorporates producibility, simplicity, standardization, modularity, flexibility, quality, and cost effectiveness for total productivity.
10. Emphasize total quality control, which begins with designing a product and process to produce a quality product that does not rely on inspection after it is produced.

The primary quantity JIT principles are:

1. Improve plant layout by designing for flow rather than function, cutting manufacturing cycle times, and emphasizing flexibility and responsiveness.
2. Avoid moving parts to unofficial queue areas.
3. Make continuous improvement toward eliminating setup time (quality principle also).
4. Reduce lot sizes and produce in small lots using a "pull" production system rather than a "push" system.
5. Reduce inventory levels to approach zero where possible because excess inventory is a waste and a cover-up for other problems and poor planning.
6. Develop a uniform plant load and avoid overloading capacity.

7. Eliminate high safety stock required because of errors in production planning, long lead times, and vendor reliability problems.
8. Part scheduling should be based on finished part due dates, not operation start dates.
9. Shop supervision should be trained to be sensitive to throughput time and delivery dates, not direct labor cost.

The concept of just-in-time manufacturing is based on a philosophy of solving and preventing problems, eliminating waste, and striving for continuous improvement. For any factory modernization effort to be successful, whether automated or not, a just-in-time approach should be considered. A commitment to JIT, just like FMS, requires management involvement, sound planning, and acute attention to detail.

SUMMARY

1. Just-in-time is a stockless production manufacturing approach whereby only the right parts, both purchased and manufactured, are usable, completed, and available at the right time and in the right place.

2. JIT is an operation philosophy and an operation strategy, not a system, and must be viewed as an opportunity, not a threat.

3. JIT is about organizing the production process so that inventory of raw materials, work in process, and finished goods are driven to a minimum and kept under control.

4. Besides reducing inventory and improving material flow, JIT can also shorten product development times, improve order entry cycle times, and improve every step of the production cycle.

5. JIT can be either a cause or an effect of FMS; however, it is difficult to implement FMS without JIT and achieve the true results of FMS.

6. FMS can be used in many cases to drive productivity improvement changes like JIT and group technology through the organization.

7. Two implementation cornerstones that are vital to just-in-time implementation success are improved material flow and setup reduction.

8. Quality-related problems in an organization should be addressed before quantity issues.

REVIEW QUESTIONS

1. JIT is:
   (a) An operation system to reduce inventory
   (b) A stockless production manufacturing approach where only the right parts are completed at the right time

      (c) A turnkey system purchased with FMS

      (d) Material flow improvement and setup reduction

2. High inventory levels are used in many cases to cover up other problems in an organization. True or False?

3. The primary goal of JIT is to have only the lowest necessary level of material arrive at the plant, or be made within the plant, exactly when it is ready to be used. True or False?

4. JIT is a "pull" system, where parts are produced only when they are needed, while a typical batch-oriented factory relies on a _____ system of large lot size production.

5. Which of these is *not* a direct benefit of just-in-time manufacturing?

      (a) Reduced departmental part moves

      (b) Reduced process planning time

      (c) Increased capacity utilization

      (d) Reduced floor space

      (e) Increased part throughput

6. Having to maintain high inventory buffer stock to compensate for variations in vendor deliveries is a sound reason not to implement JIT. True or False?

7. Explain the relationship between JIT and FMS.

8. Overall part flow is dictated by the slowest producing work station or element. True or False?

9. List three ways to reduce setup time.

10. Making every worker responsible for his or her own rework is a _____ JIT principle.

Group Technology

8

OBJECTIVES

After studying this unit, you will be able to:

- [ ] Describe group technology and discuss the importance of a coding and classification system.
- [ ] List and identify key GT benefits by functional area.
- [ ] Explain the relationship of GT to FMS.
- [ ] Name four implementation–application principles.

INTRODUCTION

Group technology, as a concept, has been around since World War II but has only recently received widespread attention due primarily to its close association with cellular manufacturing, FMS, CIM, and other factory automation programs. Group technology (GT) is not simply the formation of machinery into manufacturing cells, although cellular arrangement is a logical consequence of group technology application. It involves bringing together and organizing (grouping) common concepts, principles, problems, tasks, and technology to improve productivity. Many companies that use GT effectively have not assembled machine tools into cellular arrangements in their plants and many companies with cells are making poor use of GT.

Group technology, like JIT, is a journey, not a destination. It involves continuous improvement and structured discipline and must be a fundamental building block of a cell or system if the real benefits of automation are to be achieved. And it must be approached and applied before, during, and after automation.

Group technology begins with an overall strategic review of the business plan: goals, products, processes, resources, and applicable technologies. It should be part of the planning process when preparing for a cell or system, but, like JIT, it does not necessarily require a cell or system to be applied. Adoption of a plan and a strategy to progress toward group technology must be tailored to a company's particular needs, applicable technologies, and available resources. It requires keen understanding, careful consideration, disciplined commitment, and acute attention to detail.

WHAT IS GROUP TECHNOLOGY?

There are many definitions surrounding the term group technology, or GT for short. It has been described as a concept, a philosophy, an organizational principle, a discipline, and a method. GT is, in fact, all of these.

Broadly defined, group technology implies the grouping of various technologies to achieve a competitive edge based on a predefined operational strategy. Group technology generally implies the physical rearrangement of manufacturing from the typical job shop cluster of similar machines to the not so typical cluster of dissimilar machines into cells to increase throughput and decrease part move and queue time. Moving further upstream toward engineering, GT further implies discipline, control, and stability of part family designs, along with designs for manufacturing effectiveness through standardization of design features and part attributes within families.

Group technology is the realization that many parts have similar geometric features, and by combining those design requirements, a common part solution can be found. Parts may be grouped or arranged for group technology through:

1. Design characteristics or features
2. Manufacturing processes

Design engineers think primarily in terms of function and performance and like to be creative. Reviewing and grouping parts by design characteristics or attributes helps to minimize design duplication and creating a new part when an existing part may do the job. Creating a new part and introducing a new part number to be manufactured are very expensive. If the new part somewhat resembles a previously designed part and can be used or modified, introduction costs can be cut considerably. For example, introducing a new part number when an already existing part could be used means that a new process plan would need to be created, one or more NC programs may be required, and tooling and fixturing may need to be made or ordered.

Grouping parts by manufacturing processes is really grouping manufacturing machines into correspondingly grouped and specialized focus factory work cells instead of the traditional arrangement of machines according to function (see chapter 2). Each work cell can be specially arranged to produce a given family of parts. Grouping parts into families to be processed on machines grouped for similar manufacturing processes can considerably reduce manufacturing costs related to the number of part setups, amount of material handling, length of lead time required, and amount of in-process inventory. In grouping part families, where a choice has to be made between choosing to organize or arrange by design characteristic or manufacturing process, manufacturing needs should take precedence over design needs. This is because manufacturing costs account for such a high share of the final product cost.

Regardless of whether GT is used to minimize the proliferation of new part designs from product engineering or used to establish work cells for manufacturing families of parts, group technology in its pure form is not a "canned" solution to any specific problem. It is actually a management strategy for standardization of effort and elimination of waste caused by duplication of effort and affects all areas of a company, not just engineering or manufacturing.

CLASSIFICATION AND CODING

For parts to be grouped based on either design characteristics and features or manufacturing processes, they must be classified into predetermined categories and coded for retrieval and use. Classification and coding are computerized tools used to capture the design and manufacturing features of parts; they provide the ability to retrieve and analyze data by desired feature. This is essentially a system of arrangement, much like a zip code or the classification system used in a library.

A parts classification and coding system has come to be known as the glue of group technology, and as a means to identify the various design attributes or manufacturing features for parts grouping. Group technology, through parts classification and coding, is the communication link between the design and the manufacturing data base (Figure 8-1).

The primary objective of a classification and coding system is to develop a multidigit code number for each part that identifies its major attributes and features essential to its placement in a particular part family. The idea then is to group parts having similar design attributes into part families having common processing requirements, which is naturally followed by designating machine groupings to do the processing. Typical attributes represented by GT code numbers include part configuration (round or prismatic), further definition of shape, dimensional envelope, surface finish and tolerance requirements, workpiece material, raw material state (bar stock, casting, forging, and so on), and other necessary identification features. Any process or geometric variation that is helpful in uniquely describing a part can be coded in a GT coding system.

Although many types of commercially available or custom-developed coding and classification systems can be purchased, they typically fall into one of the following three categories:

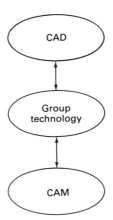

Figure 8-1   Group technology is the vital link between the design and manufacturing data base.

1. Design feature group
2. Manufacturing feature group
3. Design and manufacturing feature group

A generic example of a design feature classification and coding scheme can be seen in Figure 8-2.

Commercially available and user-specific classification and coding systems have coded part designations ranging from eight to thirty-five alphanumeric characters in length. In general, each digit has a value of zero to nine, with the first digit identifying the general part type and each succeeding digit identifying the next subgrouping. The larger and more diverse the part population, the greater the need for additional part classification digits. Most commercial classification and coding systems are designed so that the first digits are more universal in nature and become more specific and customer-definable with the remaining digits.

Parts classification and coding may take several years or more to manually code and enter part attribute data, along with a solid commitment to religiously enter coding and classification data into the computer as new parts are designed. Eventually, the data base becomes large enough that accessing the data base first to find out if a similar design already exists, for example, may yield positive results and avoid creating and making a new part.

BENEFITS AND RELATIONSHIP TO FMS

The benefits of group technology very closely parallel those of JIT, CIM, and even FMS and numerical control. And they are closely interwoven with each other. It is very difficult to implement one without affecting or seeing the benefits of the other.

The traditional payoffs of group technology have been well documented by

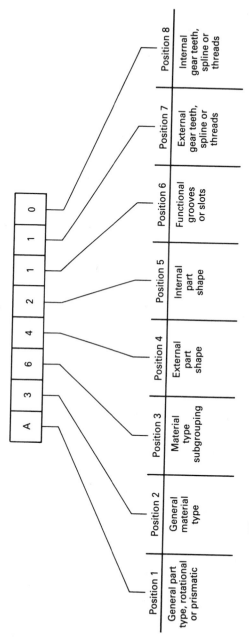

Figure 8-2   Generic example of a design feature classification and coding scheme.

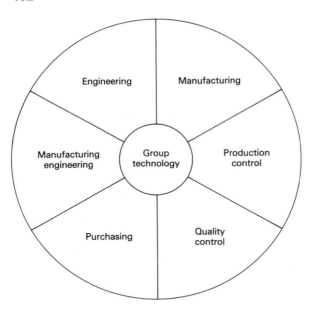

Figure 8-3   Principal areas of a company benefited by group technology.

those who have done their homework and implemented a GT program, with or without a high-tech cell or system. Savings in all areas of a company with a well-implemented GT program are typically paid back within the first two to three years. But, like JIT, CIM, FMS, or numerical control, the real benefits can only be derived if understanding, commitment, effort, and involvement are applied.

The benefits of group technology affect many areas of a company, as seen in Figure 8-3. Benefits by functional area include:

1. **Engineering Design:**
    □ Design standardization and redundancy avoidance
    □ Rapid design retrieval
    □ Reduced number of new, similar parts and elimination of duplicate parts
    □ Reduction of drafting and part detailing effort
    □ Identification of substitute parts
2. **Manufacturing:**
    □ Reduction of part setups and associated cost and time
    □ Improved estimation of machine tool requirements
    □ Improved floor space utilization
    □ Reduced material handling and transport time
    □ Improved identification and location of bottlenecked machine groups and underutilized machine tools
    □ Improved facility planning

☐ Increased use of manufacturing cells and universal production equipment

☐ Reduced need to trace and expedite parts

☐ Improved ability to handle rush orders without causing major disruptions to production

☐ Improved control and predictability of manufacturing costs

☐ Improved quality and communications

☐ Reduced throughput time

3. Manufacturing Engineering:

☐ Reduced number of process plans and process planning time

☐ Reduced number of NC programs and NC programming time

☐ Reduced producibility analysis

☐ Improved uniformity and process plan routing

☐ Reduced tools and fixtures to be used

☐ Standardization of routings

☐ Reduction in tool design and procurement

☐ Use of common tooling and/or avoidance of new tooling

4. Production Control

☐ Reduced in-process inventory

☐ Reduced inventory warehousing, material movement, and lost or misplaced parts

☐ Easier location of production difficulties

☐ Improved equipment monitoring and scheduling

☐ Tighter and improved shop scheduling

☐ Improved capacity planning and accountability

5. Quality Control:

☐ Improved opportunities for controlling quality at the source

☐ Reduced time to locate part defects

☐ Reduced sampling and inspection time

6. Purchasing:

☐ Grouping parts for quantity buys at lower cost

☐ Establishing vendor capabilities by code to build bid lists

These are just some of the more prominent benefits of group technology. However, others exist that are not so prominent or easy to quantify. For example, how can an economic price tag be placed on the value of getting a product to market 30 percent faster than the competition? Or what is the value of being able to respond to a customer's rush order with a 50 percent decrease in response time? What price tag can be placed on being able to retain manufacturing engineering knowledge with an aging and retiring manufacturing engineering work force? These are areas that cannot always be identified and quantified but are crucial to the continued, long-term productivity and profitability of a company.

The classification and coding associated with GT provides a means by which parts may be easily selected to load a cell or FMS, if the part classification and coding system is available and operational at the time cell or system planning begins. In many instances, however, this is not the case and the part selection process is done manually. When planning for a cell or system, you are essentially doing a portion of group technology simply by determining equipment requirements, deciding which parts go into the cell or FMS, and grouping them accordingly.

An FMS is, in fact, a grouping of machines to process a family of parts within a predefined range of part feature and characteristic requirements. As mentioned before, an FMS in many cases is referred to as a cell, depending on system size and how a particular company views its automation efforts. Given these considerations, it is apparent that an FMS can actually be considered a highly sophisticated GT manufacturing cell that can produce a wider range of parts and part families than the traditional GT manufacturing cell.

IMPLEMENTATION AND APPLICATION PRINCIPLES

Group technology, as we have seen, is a method of grouping together similar parts or items for more efficient treatment from design engineering to the manufacturing floor. GT implementation and application should begin with a thorough review of products, processes, and resources, and ideally, like JIT, should be in place before adding a cell or FMS.

GT planning should initially be upfront in the upstream functions of product planning and design engineering and secondarily in the downstream functions of manufacturing engineering, production control, and manufacturing (Figure 8-4). This is because manufacturing and its related support functions basically react to front-end functional decisions. However, small efforts undertaken in the upstream functions of product planning and design engineering can have considerable impact in the downstream functions of manufacturing engineering, production control, and manufacturing. The bottom line is that the effectiveness and productivity of the manufacturing operation have become more demanding with considerably less room for internal ''man-handling'' of problems on the shop floor. A greater dependence has developed on the decisions and performance of the upstream functions. Group technology is a method of organizing and controlling the upstream and downstream functions in an industrialized business and applying discipline and control, not only to administer the techniques, but to follow the rules once the rules have changed.

Successful implementation of group technology concepts depends on the understanding, development, and application of some key GT principles. In general, these principles are:

1. Investigate the purchase of a commercially available coding and classification system. Coding new and existing parts and entering the information into the engineering data base for analysis and retrieval are the most effective means of capturing upstream information for downstream usage.

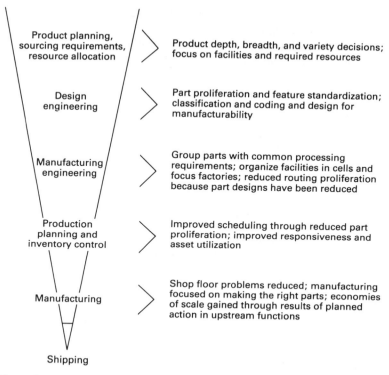

Figure 8-4 GT efforts undertaken in upstream functions can have a considerable impact on downstream functions.

2. Once a significant amount of part classification and coding data is available on the engineering data base, engineering should begin accessing the information before designing a new part to reduce new part introductions and take advantage of existing part designs if possible.

3. Design for manufacturing and assembly should be emphasized through design teams made up of select individuals from design, manufacturing, manufacturing engineering, and purchasing.

4. Eliminate engineering reward and measurement systems for "creating" new designs. Continue to encourage use of existing designs to reduce new part introductions.

5. Strive to achieve a closer match between part requirements and machine capabilities.

6. Reorganize part routings and manufacturing facilities around part families and focus factories.

7. Determine part routings based on part features (supplied by GT) and the available machine tools.

8. Apply quick setup tooling and fixturing to improve short-run capabilities and improve throughput.

9. Do not violate cell integrity by moving parts assigned for one cell to another in the name of improved efficiency, faster throughput, and increased asset utilization.

10. Control quality at the source and eliminate nonvalue-added manual inspection operations.

SUMMARY

1. Group technology is a concept, a philosophy, an organizational principle, a discipline, and a method.

2. Group technology implies discipline, control, and stability of part family designs and process plans and the physical rearrangement of the manufacturing floor.

3. Parts may be grouped or arranged for group technology through design characteristics or features and/or manufacturing processes.

4. Manufacturing needs should take precedence over design needs when grouping part families and a choice has to be made.

5. The parts classification and coding system is the glue of group technology and is a means to identify the various design attributes or manufacturing features for parts grouping.

6. The primary objective of a classification and coding system is to develop a multidigit code number for each part that identifies its major attributes and features essential to its part family placement.

7. Part classification and coding systems typically fall into the categories of (a) design feature group (b) manufacturing feature group, or (c) design and manufacturing feature group.

8. The benefits of GT very closely parallel those of JIT, CIM, FMS, and numerical control and are very closely interwoven with each other.

9. GT benefits affect many areas of a company, consisting of engineering design, manufacturing, manufacturing engineering, production control, quality control, and purchasing.

10. An FMS is a highly sophisticated GT manufacturing cell that can produce a wider range of parts and part families than the traditional GT manufacturing cell.

11. GT planning should begin in the upstream functions of product planning and design engineering and then follow in the downstream functions of manufacturing engineering, production control, and manufacturing.

REVIEW QUESTIONS

1. Physical rearrangement of the shop floor into group technology cells helps to minimize design duplication. True or False?

2. Grouping parts into families to be processed on machines grouped for similar manufacturing processes can reduce manufacturing costs related to:
   (a) Number of part setups
   (b) Amount of material handling
   (c) Length of lead time required
   (d) Amount of in-process inventory
   (e) All of these

3. GT can be considered a strategy for the standardization of effort and elimination of waste caused by duplication of effort. True or False?

4. Group technology, through parts classification and coding, is the communication link between the _____ and _____ data base.

5. Describe how a coding and classification system is established and how it can be useful in support of group technology.

6. Reduced need to trace and expedite parts is a _____ GT benefit.

7. Improved capacity planning and accountability are _____ GT benefits.

8. Explain the relationship of GT to FMS.

9. GT planning begins in the downstream functions of manufacturing engineering, production control, and manufacturing. True or False?

10. Explain how design for manufacturing and assembly is an important GT application principle.

Part III

Processing and Quality Assurance Equipment

Flexible manufacturing cells and systems continue to be implemented by users with increased emphasis being placed on computer hardware and software. Although important to overall system control and unattended operation, the computer hardware and software do not provide the essential value-added part processing within the cell or system.

The machine tool, the wash station, and the coordinate measuring machine (CMM) are the principal processing equipment. These machines provide the essential capability to machine the workpiece, clean and dry it, and inspect it during system operation. In many cases, their functionality and importance to system success is almost considered a given. However, system processing equipment, in particular the machine tools, should never be taken for granted because they form the foundation of a cell or system. Each major component must be keenly understood for the task it must accomplish and carefully selected based on a thorough analysis of the part spectrum to be produced. The accuracy, quality, repeatability, and reliability of their operation is critical to overall system performance and success.

9

Turning Centers

OBJECTIVES

After studying this unit, you will be able to:

□ Identify turning center types and describe principal design attributes and related advantages.

□ Explain primary turning center axis, programming, and format information.

□ Discuss the importance of work-holding and work-changing equipment to rotational cells and systems.

□ List and describe four automated turning center features.

INTRODUCTION

Increased and improved machine tool technology has taken the NC lathe beyond the scope of conventional turning operations. Approaching the flexibility of machining centers, a new array of NC turning center features and options has emerged that extends the turning center's capability far beyond its earlier predecessors.

Because of their advanced features and capabilities, turning centers have high utilization rates, which result in increased wear and tear on the machine and its components. Production applications range from high volume, low variety to high variety, low volume. And turning centers can accommodate a wider range of dissimilar parts in a shorter period of time with the application of new work-holding and work-changing features and enhancements.

Modern turning centers have advanced considerably in capability, flexibility, versatility, and reliability. They continue to evolve and broaden their role in rotational part automated cells and systems.

TYPES, CONSTRUCTION, AND OPERATIONS PERFORMED

The NC lathe, only partially resembling manual models, continues to be the mainstay of rotational part metal-removal machine tools. However, NC lathes today now must not only be able to handle increasing demands for flexibility, automation, and higher horsepower, speeds, and stock removal rates, but must also permit work-area access by such devices as robot loaders/unloaders, chuck-changers, and tool-changers.

Now commonly referred to as turning centers because of their increased flexibility and capability, NC lathes are classified in two types; vertical and horizontal. Vertical NC turning centers, (Figure 9-1) are modern adaptations of the manual vertical turret lathes (VTLs). Although principally designed for large and unwieldy workpieces, vertical turning centers have advanced considerably in state of the art NC machine and control unit technology. Other than enhanced features

Figure 9-1   Vertical CNC turning center. (Courtesy of DeVlieg Sundstrand)

and added technology, the basic construction of vertical NC turning centers has been, for the most part, fundamentally unchanged.

Horizontal NC turning centers of the shaft, chucker, or univeral type (Figures 9-2 and 9-3) have not only changed relative to advanced features and technology, but in basic construction as well. Many modern horizontal NC turning centers are of the slant-bed design (Figure 9-4), some with outside-diameter (OD) and inside-diameter (ID) tools mounted on the same indexable turret. Advantages of the slant-bed design include easy access for loading, unloading, and measuring, allowance for chips to fall free, minimum floor space utilization, ease and quickness of tool-changes, and better strength and rigidity. Bed designs are typically constructed of steel weldments, which generally supply greater rigidity, or are made of cast iron, which tends to dampen vibration better. Some beds are even made of reinforced concrete and epoxy-granite resin. Machine tool beds composed of this material provide excellent vibration dampening properties, along with high static and dynamic rigidity. Development continues on the use of various materials, such as steel–concrete composites and ceramics, for both horizontal and vertical turning machine bed and base construction.

Increased performance requirements have brought about considerable change in turning center headstocks and drive systems. Headstocks are either gear driven or gearless (belt driven or motorized direct drive). Geared types typically have broader speed ranges, accommodating the needs of different materials that

Figure 9-2   Horizontal CNC turning center. (Courtesy of Cincinnati Milacron)

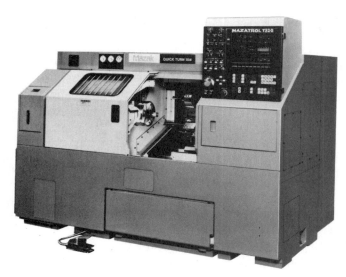

Figure 9-3 Horizontal turning centers come in a variety of types and sizes. (Courtesy of Mazak Corp.)

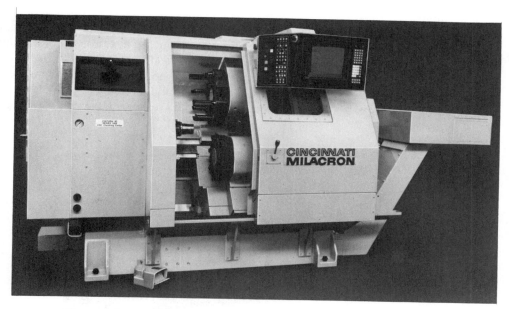

Figure 9-4 Horizontal turning center, illustrating a slant bed design. (Courtesy of Cincinnati Milacron)

require constant cutting speeds from low to high ranges. Gearless drives, which have become increasingly popular, offer the advantages of high speed and improved surface finish and accuracy as a result of eliminated gears, clutches, and shafts. Also, lower costs are the result of the elimination of these parts. However, the choice of drive depends on the application, and the user must weigh such variables as workpiece size and material to determine which type best suits overall requirements.

Four-axis and dual-spindle turning centers (Figures 9-5 and 9-6, respectively) have also gained considerable acceptance among manufacturers. Four-axis lathes provide considerable savings and productivity increases over conventional two-axis machines because OD and ID operations can be performed simultaneously through independently programmed slides. Dual-spindle machines have two spindles with two independent slide motions for OD and ID operations for each spindle. Machines of this type are also capable of achieving high productivity levels with considerable savings.

Turning center movements are controlled through the CNC unit. Programming is typically accomplished off-line by an NC programmer via computer or manual means, or on the shop floor by the operator through sophisticated CNC interactive graphics systems similar to that depicted in Figure 9-7. Such shop floor programming controls are geometry, process, or feature–motion oriented and allow the part to be programmed as it is shown on the part print. Foreground–background processing permits programming of a new part while another part is being cut. Other unique shop floor programming features such as menu-selectable tool data and extensive diagnostics provide a high level of user-friendly operator interface.

Operations performed on NC turning centers are basically no different than those performed on older or conventional machines. These consist of the standard turning, facing, drilling, boring, tapping, and threading (Figures 9-8 and 9-9). These modern CNC versions can still remove metal no faster than their conventional counterparts; however, enhanced cutting tool technology, tool-changing and work-holding methodology, and new automated features have considerably short-

Figure 9-5   Four-axis CNC turning center. (Courtesy of Mazak Corp.)

Figure 9-6 Dual-spindle vertical CNC turning center. (Courtesy of Olafsson Corp.)

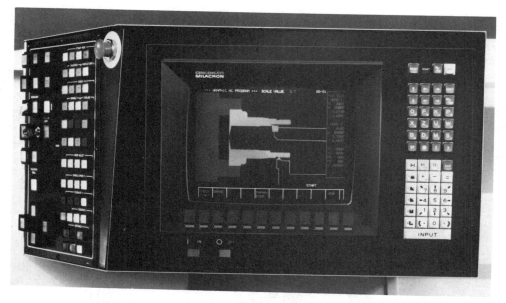

Figure 9-7 CNC interactive graphics control unit. (Courtesy of Cincinnati Milacron)

Figure 9-8   Turning operation on a VTL. (Courtesy of DeVlieg Sundstrand)

Figure 9-9   Turning operation on a horizontal CNC lathe. (Courtesy of Kennametal, Inc.)

ened the noncutting time. And added new capabilities, such as live rotary spindle tooling to perform off-center face and OD milling, drilling, and tapping, add significantly to modern turning center operational flexibility and capability.

AXES, PROGRAMMING, AND FORMAT INFORMATION

The primary axes of a turning center (Figure 9-10) are Z and X. The Z axis travels parallel to the machine spindle, while the X axis travels perpendicular to the machine spindle. U and W axes, also shown in Figure 9-10, are typically auxiliary axes providing additional movement and tool capacity. For both OD and ID operations, a negative Z $(-Z)$ is a movement of the saddle toward the headstock. A positive Z $(+Z)$ is a movement of the saddle away from the headstock. A negative X $(-X)$ moves the cross slide toward the spindle center-line, and a positive X $(+X)$ moves the cross slide away from the spindle center-line.

For positioning turning center axes, both absolute and incremental programming are used. Most older turning centers were limited to incremental positioning only; however, modern turning centers can be programmed in absolute or incre-

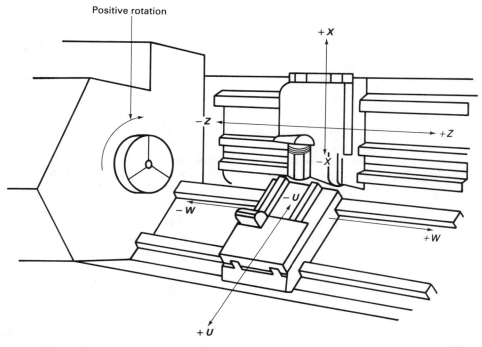

Figure 9-10 **The primary turning center axes are Z and X. U and W are auxiliary axes that provide additional movement and tool capacity on some machines.**

mental. Even when incremental positioning and programming are used, the program manuscript usually contains two extra columns (Z and X) for the programmer to keep track of the absolute dimensions from the zero point. Absolute programming, in contrast, simplifies the effort involved in programming and positioning verification, thereby helping to ensure accuracy.

Absolute and incremental programming are equally effective. For modern turning centers the choice is up to the NC programmer. If part print dimensions are given incrementally, the programmer programs a G91 for incremental and the system readies itself for incremental input. If workpieces are dimensioned in absolute form, the programmer uses a G90 for absolute input. Many CNC units for turning applications have either a G90 or G91 set as the default condition.

Programming commands for turning centers consist of linear and circular interpolation. The G01 linear interpolation preparatory function commands the slides to move the tool in a straight line from the current position to the command position. The rate of traverse is measured along the vector connecting the two points and is used for turning, drilling, boring straight diameters, facing shoulders, and turning or boring chamfers and tapers.

Circular interpolation on an NC turning center moves the tool in a circular arc along a path generated by the control system. The rate of travel is the same around the arc, with a tangential vector feed equal to the programmed feed rate. Circular interpolation is specified by a G02 preparatory function for clockwise direction and G03 for counterclockwise direction. Coordinate information is also programmed to define the start point, the end point, and the center point ($I = X$ coordinate value; $K = Z$ coordinate value). A list of all EIA and AIA standard preparatory functions (G codes) and miscellaneous functions (M codes) is provided in Appendix C. Most CNC controls for modern turning centers accept the word address, tab ignore, and variable block length tape format with either EIA (Electronics Industries Association) or ASCII (American Standard Code for Information Interchange) coding. Decimal point programming has also come into wide prominence and acceptance. Modern CNCs will automatically sense the method used (EIA or ASCII) and will decode tape image data accordingly.

The following words and tape format are used for a typical NC turning center. However, all NC and CNC systems are not the same; tape formats, words, and word addresses may differ somewhat among machine and control manufacturers.

O/N	The sequence number is composed of four digits preceded by the letter O or N (O$\times\times\times\times$ or N$\times\times\times\times$). This word is used to indicate the block of information that is being processed by the control.
G	The preparatory function code is a two-digit number preceded by the letter G (G$\times\times$). These codes are used throughout the program to define the various modes of operation.
X/Z	Axis dimensions are used to denote the position of the axes. The axes are addressed with a seven-digit number preceded by the letter X or Z. The sign denotes the direction of travel, using

the incremental mode, and the position relative to program zero, using the absolute mode (X ×××.××××).

I/K   Center point coordinates are used to define the center location for programming circular arcs. Center point coordinates are addressed with a seven-digit number preceded by a plus (+) or minus (−) sign and the letter I or K. The center point coordinates can be either absolute or incremental, depending on the input mode. I represents the X axis, and K represents the Z axis (I/K ×××.××××).

I/K   The axis feed rate for threading is controlled by programming a lead value. This is normally a seven-digit, unsigned number preceded by the letter I or K. Values for thread lead are not affected by the absolute or incremental input modes. The lead values are always positive, and the sign is not programmed. Programming a negative value will usually result in a program error. I represents X-axis lead, and K represents Z-axis lead (I/K××.×××××).

F   The axis feed rate is controlled by a four-digit number preceded by the letter F. Feed rates may be programmed in either distance or travel per minute or distance per revolution of the spindle, depending on the selected preparatory function (F×××.× −IPM or F.×××× −IPR).

R   The radius dimension used for CSS (constant surface speed) programming is a seven-digit number preceded by a plus (+) or minus (−) sign and the letter R. The R dimension is always an incremental value measured from the spindle centerline to the tool tip (R ×××.××××).

V   The tool retract feature is programmed with a two-digit V word and programs a tool retraction along an interference-free path. The two digits of the V word represent the X and Z axes, respectively. The value of the digit that is programmed determines the direction and distance the tool will travel when the operator initiates the tool retract feature (V××).

S   Spindle speeds are programmed with a four-digit number preceded by the letter S. The spindle speed can be programmed in either direct rpm coding or in the CSS mode (S××××).

T   The turret station and offsets are programmed with a four-digit number preceded by the letter T. The first and second digits usually identify the turret station, and the third and fourth digits represent the offset. (T××××).

C   The C word is used to define the total number of thread starts and the thread start to be machined when machining multiple-start threads.

D   The taper trim feature compensates for taper in the workpiece and is programmed using a two-digit D word. The D word designates the number of the active taper trim pair. The operator must manually input the compensation values. Both leading and trailing zeros must be programmed. The D word uses the same format for both inch and metric values $(D \times \times)$.

M   The miscellaneous function codes are two-digit codes preceded by the letter M. These codes are used throughout the program to perform functions such as spindle starting and stopping, coolant control, and transmission range selection $(M \times \times)$.

WORK-HOLDING AND WORK-CHANGING EQUIPMENT

Improvements in work-holding and work-changing equipment have advanced considerably, keeping pace with the rapid technological development of modern turning centers. This equipment, consisting of automatic chuck-changing systems, countercentrifugal chucks, automatic pallet-changing systems, and automatic chuck-jaw changing systems, enhances the productivity of rotational part cells and systems. Increased machine utilization, coupled with demanding production and automation requirements, has created this demand from both machine tool builders and users for high-performance work-holding systems.

Rotational work-holding equipment, most commonly chucks and special application face plate drivers and chucks (Figures 9-11 and 9-12), must safely hold the workpiece to withstand the tremendous cutting forces generated by the metal-removal process. It must also compensate for the centrifugal forces that, at high rotating speeds, counteract the clamping power provided by the work-holding equipment with a tendency to open the chuck. This constant and dangerous jaw force loss at high speeds can be nearly eliminated with a centrifugally compensated chuck. Countercentrifugal chucks (Figure 9-13), which are available from a variety of manufacturers in wedge, lever, and wedge/lever actuated types, incorporate counterweights pivoted so that centrifugal force tends to increase the grip, offsetting the outward developed centrifugal jaw forces.

A variety of other actuated chucks (Figures 9-14 on page 182 and 9-15 on page 183) is available, consisting of through-bore power chucks, draw-down chucks, and retractable jaw chucks. All commonly available standard and power chucks provide radial adjustment of the jaws to clamp parts having different diameters. For OD and ID clamping, the universal top jaws are reversible and contain several steps for different diameter ranges. These offer a wide variety of automated work-holding and clamping capabilities for rotational cells and systems.

Achieving higher speeds and greater metal-removal rates are not the only concerns with NC turning centers. Equally important to increasing the up time of these machines for short and medium runs are chucks and chuck jaws that can be quickly changed over. Formerly, changing chuck jaws could take 20 to 30 minutes. But new quick-change designs allow jaws, chucks, or pallets to be changed in less

Figure 9-11   Chuck for a modern NC turning center. (Courtesy of Cushman Industries)

Figure 9-12   Modern work-holding equipment enhances the capabilities of rotational cells and systems. (Courtesy of Cushman Industries)

Figure 9-13 Countercentrifugal chucks incorporate internal counterweights that pivot to offset centrifugal force when the chuck rotates, thereby increasing the grip on the workpiece. (Courtesy of ITW Woodworth)

than 2 minutes. The smaller the lot size is the more chuck, jaw, or pallet changing that is required in order to provide quick changeover to a new lot of parts and to keep the machine or machines in production. The introduction of automatic tool changing systems, robot-controlled workpiece material handling, and electronic sensing led to the development of automatic jaw-changing systems. Various jaw-

Figure 9-14 Power chuck capabilities have advanced considerably, keeping pace with modern turning technology. (Courtesy of ITW Woodworth)

Figure 9-15   Power chucks are available in a variety of types and sizes. Illustrated here is an example of a wheel power chuck. (Courtesy of Cushman Industries)

changing systems are available and they are an important factor in the operation of an automated rotational cell or FMS.

Modern jaw-changing systems (Figure 9-16) are controlled hydraulically, but generally have an electronic stroke control to select, set, and secure the desired clamping range and position. With this type of system, a 12-inch-diameter chuck, utilizing five sets of manual quick-change jaws, has a flexible clamping range from approximately 1 to 9 inches in diameter. In this type of environment, automatic changing of work-holding tooling and components further reduces human interference and provides the flexibility for short-run production without extensive and expensive machine downtime during changeovers.

In some applications, workpiece variety may not permit an automated jaw change because too great a size variation exists between different workpieces. In such cases, the chuck itself may be changed, as seen in Figure 9-17. This can be accomplished by a hydraulic locking–releasing mechanism or by a robot. In some cases, a workpiece held in a manually operated chuck can be loaded and unloaded outside the machine work area. At the end of each complete machining cycle, an automatic manipulator can replace the chuck holding a completed part with a chuck holding a new part to be machined. Because part loading and unloading are completed outside the work area and internal to the machine cycle (while the machine is in production), machine downtime can be almost eliminated.

Figure 9-16 Jaw-changing systems are hydraulically controlled with an electronic stroke device to select, set, and secure the desired clamping range and position. (Courtesy of Mazak Corp.)

Figure 9-17 Chuck-changing systems are used in applications where workpiece size variation is too great for automated jaw changing. (Courtesy of ITW Woodworth)

Pallet-changing systems (Figure 9-18) are increasingly being utilized for the machining of complex parts. These typically require less than 1 minute for changeover and employ the use of a manipulator that rotates the pallet into the vertical plane and positions the pallet precisely on the turning center's spindle. Hydraulic clamp fingers are actuated and securely lock the pallet to the spindle face. The concept expands flexibility by allowing the entire pallet and part assembly to be moved to other machines within a cell for drilling, milling, or tapping operations.

Countercentrifugal chucks, automated jaw-changing equipment, and chuck- and pallet-changing systems are just some of the work-holding systems being used to broaden the scope and flexibility of automated turning cells and systems. Work continues on the development of better and more automated methods to enhance and improve work-holding and work-changing equipment for rotational part applications.

Figure 9-18 **Automated pallet-changing systems can position a pallet with a new part on a turning center spindle in less than 1 minute for complete changeover. (Courtesy of Cincinnati Milacron)**

AUTOMATED FEATURES AND CAPABILITIES

New turning center features and capabilities have taken original concepts of the NC lathe into a new dimension. The ability to perform multiple operations on a workpiece and run unattended provide a new degree of flexibility and greater compatibility with other components of an automated rotational part system.

Automated features and capabilities perform a multitude of functions that now automate what was previously performed manually by skilled operators. The principal automated turning center features and capabilities are:

1. Automatic Gaging

The two broad categories of automated gaging are in-process and postprocess gaging. In-process gaging is generally considered best because it is a technique of actively measuring the workpiece and adjusting for size as it is being cut. This process is frequently performed in grinding operations but is not very practical for turning because of the adverse metal-removal conditions involved. Postprocess gaging is primarily used for turning applications including off-machine gaging and on-machine touch-trigger probing (Figure 9-19). Off-machine gaging is generally preferred if the part is inexpensive, has short cycle times, and is run in large quantities. On-machine touch-trigger probe gaging is advantageous for small batch runs where gaging flexibility is important or when machining times are long and the extra time consumed for probing minimizes part throughput interference.

Figure 9-19 Touch-trigger probing is a widely used example of postprocess, on-machine gaging. (Courtesy of Mazak Corp.)

2. Probing

Sometimes referred to as tough-trigger probing, the probe (Figure 9-20) is a very effective measuring device having applications in a wide variety of situations. The probe is a high-precision switch. The probe stylus is attached to the switch, and when the stylus is deflected (Figure 9-21), the switch contacts close, allowing current to pass. The machine control must be equipped with the proper software, which continually scans the probe's input; and when a contact closure is detected, it captures the current values in the machine's active position register. The values from the active position registers can be compared to the expected values in the part program to calculate a deviation. This deviation can then be entered as an offset to correct or regrid the machine axes. Probing accuracy depends on the repeatability of the switch, the accuracy of the values in the machine position registers, and the speed at which the control can access the values in the registers. Accuracy can be increased by automatically recalibrating the probe by touching it to a reference surface, usually the machine's chuck.

3. Live Spindle Tooling

Rotary tooling (Figure 9-22) is one of the most important new features associated with advanced turning centers. Available on both horizontal and vertical

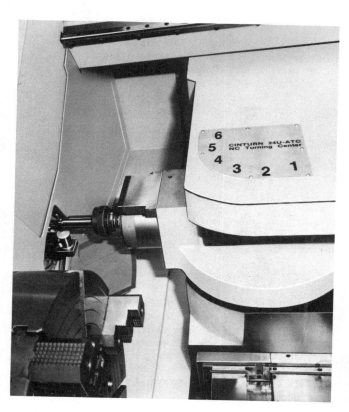

Figure 9-20 Probing is an effective measuring method with a wide variety of applications and uses. (Courtesy of Cincinnati Milacron)

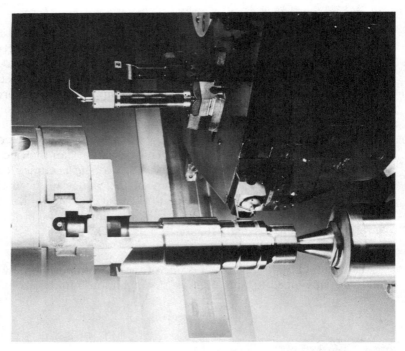

Figure 9-21   The probe is a high-precision switch whose accuracy depends on the repeatability of the switch, stored machine register values, and control register accessing speed. (Courtesy of Cincinnati Milacron)

Figure 9-22   Live spindle tooling is one of the most important new turning center features because it permits rotational tool operations on the part while the workpiece remains clamped in the chuck. (Courtesy of Cincinnati Milacron)

Figure 9-23 Automated turning center tool-changing systems are available from those that change only the cutting tool heads to those that change the entire tool holder. (Courtesy of Cincinnati Milacron)

machines, it permits milling, drilling, and tapping on the part face and outside diameter while the workpiece remains stationary and clamped in the chuck or on a face plate. Operations that can be performed with rotary tools can be divided into two broad categories: those requiring only a spindle index to an oriented stop, and those in which spindle rotation (C axis) is a feed motion moving simultaneously with X, Z, or both linear axes. Separate drilling and milling machines for secondary operations on turned workpieces, along with the additional piece handling time, can be eliminated in many cases. And accuracy is improved since milling and drilling are done in the same setup, eliminating refixturing and its effects on concentricity. However, the addition of milling, drilling, reaming, and tapping cycles allows the completion of workpieces on the turning center at a cost of increased cycle and programming times. Consequently, live spindle turning machines are most effective for small to medium lot sizes.

4. Tool Change Systems

The typical NC turning center's twelve-to sixteen-tool turret cannot provide enough tools and tool inserts for flexible unattended turning operations. Tool-changers on NC turning centers have been available since the early 1970s, with completely automated tool handling becoming a reality in the 1980s. Tool-changing systems vary from those that change only the cutting tool heads (Figure 9-23) to those that change the entire toolholder, where tools reside in a tool storage drum or magazine (Figure 9-24). In some cases, the tool assemblies are similar to and interchangeable with machining center tool assemblies. Many of these tool-changing and tool-storage systems are located at one side of the machine, and some vertical turning centers swap tools to the ram either by moving the ram laterally to the tool drum or magazine or by removing the tool and transporting it to the ram. The principal benefits of automated turning center tool-change systems are:

1. Reduction in setup time
2. Reduction in insert changing time
3. Added machine flexibility
4. Improved accuracy and reduced errors through a reduction of manual intervention
5. Increased machine tool utilization

Figure 9-24   On some automated turning centers, tool capacity is increased through the addition of a tool storage drum or magazine. (Courtesy of Cincinnati Milacron)

Figure 9-25 Tool monitoring and sensing help to considerably reduce operator attention to cutting tool wear and breakage conditions. (Courtesy of Kennametal, Inc.)

5. Tool Monitoring and Sensing

Tool wear and breakage (Figure 9-25) require operator attention as often as every 3 minutes in turning operations. Tool-monitoring and tool-sensing systems substitute for the skilled operator's eyes and ears and signal the need to replace worn or damaged tools. The simplest form of tool-monitoring is not a true tool-monitoring system. It simply records the actual cut time of each tool, compares that with preprogrammed limits for the tool's life, and signals for a tool-change when the programmed limit has been reached. Other types of tool wear systems include:

1. *Horsepower sensing:* Based on the fact that a worn tool draws more power than a sharp tool, this type of tool wear system measures the load on the main spindle drive motor. If the horsepower exceeds the programmed limits, indicating a worn tool, a tool change is signaled.

2. *Acoustic-emission sensing:* This tool wear–breakage subsystem operates on the principle that cutting processes emit acoustic pulses at ultrasonic frequencies. When a tool is about to break, acoustic emissions increase up to five times their normal value. When the acoustic-emission sensor detects a rapid increase in acoustic emissions as compared with preprogrammed values, the feed can be stopped and a tool-change initiated.

3. *Learn-mode systems:* This is a feed-force monitoring system that memorizes the sensor signal values from sharp tools. If a predetermined percentage

increase of the feed force is exceeded, a tool change to replace the worn tool at the end of the operation is initiated. In the event of tool breakage, the monitor instantly senses the sudden force increase and signals the control to stop the feed and change to a new tool.

4. *Force-monitoring:* These typically monitor the force on ball-screw drives in the X and Z axes. Sensors that support X and Z axis ball-screw drives and bearings measure fluctuations in the feed force and, when excessive up to a predetermined level, initiate a tool change.

SUMMARY

1. NC turning centers are classified in two types: vertical and horizontal.

2. Many NC turning centers are of the slant-bed design, offering the advantages of easy loading, unloading, and measuring, allowance for chips to fall free, minimum floor space utilization, and better strength and rigidity.

3. Turning center headstock drive systems are typically of the geared or gearless type; gearless types have become increasingly popular and offer the advantages of higher speed, improved surface finish and accuracy, and fewer parts.

4. Considerable advances have been made in turning center CNC units, permitting shop floor programming through sophisticated CNC interactive graphics.

5. The primary turning center axes are Z and X; the Z axis travels parallel to the machine spindle, while the X axis travels perpendicular to the machine spindle.

6. Absolute and incremental programming and positioning are both used on modern turning centers.

7. Most CNC units for turning centers use either EIA (BCD) or ASCII coding and conform to EIA and AIA standards.

8. Work-holding and work-changing equipment, consisting of automatic chuck-changing systems, countercentrifugal chucks, automatic pallet-changing systems, and automatic chuck- and jaw-changing systems, add considerably to the productivity of rotational cells and systems.

9. Principal automated turning center features include automatic gaging, touch-trigger probing, live spindle tooling, tool-changing systems, and tool monitoring and sensing.

REVIEW QUESTIONS

1. The two types of NC turning centers are _____ and _____ .

2. List the advantages of a slant-bed design for NC turning centers.

3. Some NC turning center beds are made of reinforced concrete and epoxy-granite resin because they weigh less than steel. True or False?

4. Gearless headstock drives:
   (a) Offer higher speeds than gear-type drives
   (b) Have fewer headstock parts
   (c) Are more expensive than gear-type drives
   (d) Only (a) and (b)
   (e) All of the above

5. A negative $X(-X)$ axis move on an NC turning center is a move of the cross-slide away from the spindle center line. True or False?

6. The two types of NC turning center programming and positioning are _____ and _____ .

7. Programming commands for turning centers consist of linear and circular interpolation. True or False?

8. The two NC word addresses that are used to define center points for circular axes are _____ and _____ .

9. Briefly describe how a countercentrifugal chuck functions.

10. Automatic chuck-changing systems are used because, in many cases, too great a size variation exists between different workpieces for jaw changing. True or False?

11. List four principal automated turning center features.

12. An acoustic-emission sensor is a feed-force monitoring system. True or False?

13. The two types of automated gaging are _____ and _____ , of which _____ is generally considered better.

10

Machining Centers

OBJECTIVES

After studying this unit, you will be able to:

☐ Identify machining center types and list some advantages and disadvantages of each.

☐ Explain primary machining center axes, principles of programming, and format information.

☐ Name three tombstone part mounting schemes and two programming options for machining center part processing.

☐ List and describe four automated machining center features.

INTRODUCTION

Known in the 1960s as ATCs, or automatic tool changers, machining centers originated out of their capability to perform a variety of machining operations on a workpiece by changing their own cutting tools. Thus began a tool change and additional feature/capability revolution among machine tool builders that continues to escalate by adding improvements and enhancements to the staggering array of machining center choices.

Machining centers, even in the early years of "turret drills," began to affect manufacturing operations. Their adoption, in many cases, served as a shop's introduction to numerical control. And their high productivity frequently has forced a rethinking of part setup and processing requirements. Statistically, they have become well accepted as a separate class of machine.

Today, machining center usage continues to expand from stand-alone job shop applications to flexible manufacturing cells and systems. However, the increased utilization and higher chip-removal rates of automated applications place considerable wear, tear, stress, and strain on a cell or system's most value-added component.

TYPES, CONSTRUCTION, AND OPERATIONS PERFORMED

Machining centers, just like turning centers, are classified as either vertical or horizontal. Vertical machining centers (Figure 10-1) continue to be widely accepted and used, primarily for flat parts and where three-axis machining is required on a single part face such as in mold and die work. Horizontal machining centers

Figure 10-1 Vertical machining center. (Courtesy of Cincinnati Milacron)

(Figures 10-2 and 10-3) are also widely accepted and used, particularly with large, boxy, and heavy parts and because they lend themselves to easy and accessible pallet shuttle transfer when used in a cell or FMS application.

Selection of either a vertical or horizontal machining center mainly depends on the part type, size, weight, application, and, in many cases, personal preference. Each has its own specific advantages and disadvantages, as seen in the chart in Figure 10-4.

Machining center innovations and developments have brought about the following improvements:

- Improved flexibility and reliability
- Increased feeds, speeds, and overall machine construction and rigidity
- Reduced loading, tool-changing, and other noncutting time
- Greater MCU (machine control unit) capability and compatibility with systems
- Reduced operator involvement
- Improved safety features and less noise

These improvements, driven by increased productivity demands and intense

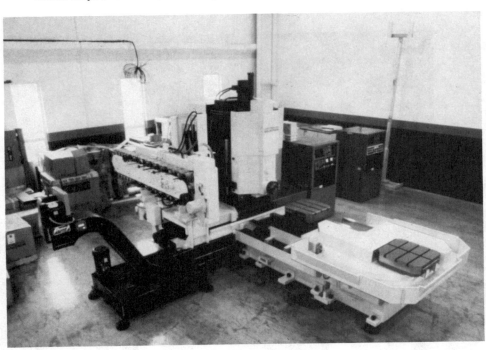

Figure 10-2   Horizontal machining center. (Courtesy of Cincinnati Milacron)

competition among machine tool suppliers, have created part-hungry machine tools able to machine workpieces to exacting tolerances accurately and consistently.

Machining center construction has improved to accommodate higher spindle speeds, feeds, and horsepower requirements, along with overall higher utilization rates and increased performance requirements. Machine beds, for the most part, are still the more traditional cast-iron or welded-steel plating. However, computer modeling of the final structures, using such techniques as finite-element analysis, has become more widespread. As a result, castings now routinely feature internal ribs and fabricated steel shapes and braces that have been optimized to yield fewer distortions when the machine is in a load or cut condition.

Spindle head improvements have advanced to accommodate a fifth axis (and more) of movement (Figure 10-5), which greatly enhances a machining center's versatility. Key to the new designs are pitch and roll motions right in the spindle head. The additional axes of movement are a necessity, particularly on larger machining centers, where the workpiece cannot be easily moved and the tool must tilt and pivot to machine the stationary clamped part. But spindle heads that move in more than three primary axes are also becoming more popular for smaller workpieces. Parts previously machined in several settings on ram-type universal

Figure 10-3   Horizontal machining centers are used in many cell and FMS applications. (Courtesy of Cincinnati Milacron)

VERTICAL		HORIZONTAL	
Advantages	*Disadvantages*	*Advantages*	*Disadvantages*
□ Thrust is absorbed directly into the machine table during deep tool thrust operations such as drilling or pocketing □ Ideal for large, flat plate work and single-surface, three-axis contouring □ Heavy tools can be used without concern about deflection □ Generally less costly	□ As workpiece size increases it becomes more difficult to conveniently look down into the cut □ Extensive chip buildup obstructs view of the gut and recuts chips □ On large verticals, head weights and distance from the column can cause head drop, loss of accuracy, and chatter □ Not suitable for large, boxy, heavy parts	□ Table indexing capability enables multiple sides of a workpiece to be machined in one setting and clamping □ Chips drop out of the way during machining, providing an uncluttered view of the cut and preventing recutting of chips □ Operator's station is to one side of the column, providing good line of sight control □ Pallet shuttle/ exchange mechanisms are open, accessible, and easy to service □ Ideally suited for large, boxy heavy parts □ Overall, more flexible	□ Difficult to load and unload large, flat, plate-type workpieces □ High thrust must be absorbed by tombstones, fixtures, or right-angle braces □ Heavy tools can deflect □ Generally more costly

Figure 10-4 General advantages and disadvantages of vertical and horizontal machining centers.

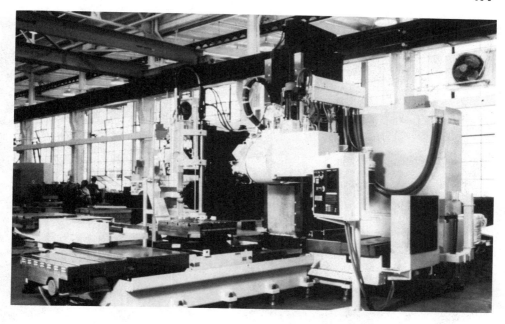

Figure 10-5   A machining center's versatility is greatly enhanced through the addition of a fifth-axis rotating spindle head. (Courtesy of Cincinnati Milacron)

milling machines are moving over to small machining centers to be completed in one part setting.

Some machine tool vendors now offer horizontal–vertical spindles (Figure 10-6). These are similar in appearance to right-angle spindle attachments that have long been available to change the spindle orientation by 90 degrees. Machines such as these continue to gain acceptance and prominence because they decrease the nonvalue-added setup and piece-handling time and increase the value-added chip-making time. Overall, there is an apparent trend to adding more features, other than just rotation for machining, at the spindle. This means that heads generally are becoming more complicated.

Characteristics demanded of machining center spindles by modern high-performance cutting tool materials include stiffness, running accuracy (runout), axial load-carrying capacity, thermal stability, and axial freedom for thermal expansion. Most importantly though, the demand is for speed. In some cases, spindle speeds have exceeded the 6000 to 7000 rpm range, depending on the manufacturer and application required.

By today's definition, a machining center must include an automatic tool changer. Tool-storage and tool-change mechanisms vary among the diversified

Figure 10-6 Machining center versatility and capability are enhanced by some vendors offering machines that have both vertical and horizontal spindles. (Courtesy of Maho Machine Tool Corp.)

machine tool suppliers, as some are front, side, or top mounted (Figures 10-7 and 10-8). The advantages of having tools stored away from the working spindle include less contamination from flying chips and better protection for an operator changing tools during machining. The double-ended, 180-degree indexing arm (also Figure 10-8) continues to be the most popular approach, although various designs of the tool gripping and clamping mechanisms will vary among builders.

More cutting tools are needed with modern machining centers, which means more tool storage capacity is required. Machining requirements for cells and systems demand that backup tools be available on-line to replace a broken tool or a worn-out tool before it breaks. Tools stored at machining centers fit into individual pockets of a machine tool's magazine or tool matrix (Figure 10-9 on page 202). Pocket designs vary, ranging from simple holes cut into a disk-shaped carrousel to individually machined pockets assembled into a chain to interconnected plastic pockets.

Both random and sequential tool selection are used on machining centers, although random is by far the most predominant type used on modern machining centers. Random tooling refers to the capability of selecting any tool from the machine tool matrix at any time. With advanced machine tool and control capabilities, cutting tools may be selected in any order, used more than once, and placed back in a different pocket from which they were originally selected, and the CNC unit will remember their new location.

Figure 10-7   Top-mounted tool-storage matrix. (Courtesy of Cincinnati Milacron)

Figure 10-8   Side-mounted tool-storage matrix with a double-ended indexing tool change arm. (Courtesy of Cincinnati Milacron)

Figure 10-9   Cutting tools are stored in a machining center's tool matrix. (Courtesy of Cincinnati Milacron)

        Sequential tooling means that tools are accessed and used in a particular order of succession. Sequential tool selection requires tools to be loaded in the exact order they will be used in the program. When the program begins, the tools are selected and used one after the other, maintaining the established sequence. The correct sequential loading of the tools is of primary importance to the operator for the successful execution of the part program.

        CNC unit advances relative to machining centers have been just as rapid as for turning centers. Control advances such as sophisticated CNC interactive graphics with foreground–background displays and multitasking software (Figure 10-10) have taken machining center controls beyond the scope of mere multiaxis contouring capabilities. The addition of increased hardware and software to modern CNC units for machining centers has permitted a higher degree of functionality and feature addition at the local CNC level.

        A variety of cutting tools is used on machining centers to perform a multitude of machining operations (Figure 10-11 on page 204). Operations range from conventional drilling, tapping, reaming, and boring to extended use of end mills and face mills for contouring and profiling cuts. High-technology carbide-insert cutting tools (Figure 10-12 on page 204) are gaining wide acceptance and prominence due to their inherent accuracies and ease of maintainability (the inserts can easily be indexed to a new cutting edge or replaced). This eliminates removing the entire tool from the holder and resharpening. Machining centers by their very design are

Figure 10-10   New CNC features for machining centers include user-friendly foreground–background displays with sophisticated multitasking software. (Courtesy of Remington Arms)

capable of bringing more cutting tools to the workpiece with more accuracy, horsepower, and rigidity than conventional machines. Consequently, more perishable tools and inserts are used up in a given period of time during machining center part processing.

AXES AND FORMAT INFORMATION

The primary axes of both vertical and horizontal machining centers are X, Y, and Z, as shown in Figures 10-13a and 10-13b on page 205 respectively. Generally, on vertical machining centers, the X axis provides the longitudinal table travel, the Y axis provides in and out saddle movement, and the Z axis provides up and down movement of the head or spindle.

X-axis movement on horizontal machining centers is also through the longitudinal table travel. Y-axis movement is up and down, provided through movement of the machine tool's knee or spindle carrier. Z-axis positioning is through in and out movement of the machine tool's saddle, table, or spindle carrier.

Figure 10-11   Multiple cutting tools are used for a variety of operations on machining centers. (Courtesy of Remington Arms)

Figure 10-12   High-technology carbide insert cutting tools operate at substantially higher speeds and feeds than conventional tooling. (Courtesy of Kennametal, Inc.)

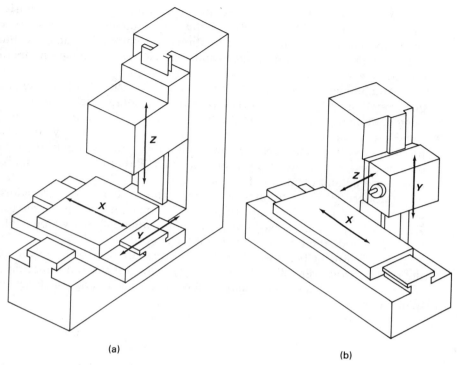

(a)   (b)

Figure 10-13   (a) Primary axes of a vertical machining center. (b) Primary axes of a horizontal machining center.

Horizontal machining centers also provide *B*- or beta-axis movement, which greatly increases a horizontal's versatility. *B* axis can be used to index a workpiece for machining in the *X*, *Y*, and *Z* plane or simultaneously with one or more other axes for a contouring cut.

There are two basic types of rotary tables that are controlled and positioned through the *B* axis. The first type of rotary table uses a positive, serrated plate to position the table mechanically. This type will lift before indexing and lower into its position after indexing. The second type of rotary table uses a rotary inductosyn seal to position and provide feed rate control. This type of table may be combined with *X*, *Y*, *Z*, and auxiliary slides to provide four- or five-axis contouring capabilities.

Axis movements of *A* and *C* are also used on some machining centers to provide rotational movement around the *X* and *Z* axes, respectively, the way *B* axis rotates around *Y*. *U* and *W* axes on machining centers are secondary linear axes to *X* and *Z*, respectively, similar to their auxiliary use on turning centers. While the addition of these axes movements increases a machining center's versatility and capability, it also adds considerable cost and compounds the part programming time and complexity.

Most rotary tables are bidirectional and will index using the shortest path to any of 72, 360, or 720 positions. Some rotary tables can index up to 360,000 positions. The different degrees of rotation in a rotary index table are illustrated in Figure 10-14. The input is in degrees and, in most cases, all positions are in absolute positioning.

Machining centers, like turning centers, accept absolute and incremental programming and EIA or ASCII input. Both absolute and incremental programming are equally effective, and the two may be used at different times within the same program. If the part print is dimensioned incrementally, the programmer programs a G91, and the CNC readies itself for incremental input. If the part print is dimensioned in absolute, a G90 is programmed. Either incremental or absolute can be set as a default condition if neither a G91 nor a G90 is programmed.

Modern CNC units for machining centers accept the word address, interchangeable-variable block tape format, or decimal point programming with either EIA or ASCII coding. Most controls will automatically sense which method is used and will decode the tape or DNC input file accordingly.

The following basic words and tape format are used on a typical NC machining center. However, all NC and CNC systems are not the same; tape formats, words, and word addresses may differ somewhat among machine and control manufacturers.

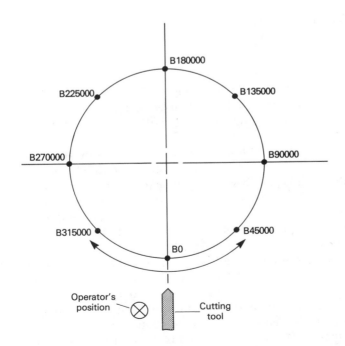

Figure 10-14   Rotary index table illustrating degrees of rotation.

O/N   Sequence number coding, introduced by O or N and generally a five-character code: one letter and four numerals ($O\times\times\times\times$, or $N\times\times\times\times$). It is used to identify a block of information. It is informational rather than functional.

G   Preparatory function coding, introduced by G. It is a three-character code: one letter and two numerals ($G\times\times$). It is used for control of the machine. It is a command that determines the mode of operation of the system. This word is informational and functional. Therefore, all characters, except leading zeros, must be included in the code.

X   X axis coordinate information code, introduced by X. It may contain up to nine characters: one letter, one sign, and up to seven numerals ($X\ \times\times\times\times.\times\times\times$ millimeters). Six numerals are normally used. This word is used to control the direction of the table travel and position.

Y Y-axis coordinate information coding, introduced by Y. This is identical to the X word, only using the Y address.

Z Z-axis coordinate information coding, introduced by Z. This is identical to the X word, only using the Z address.

R Z-axis coordinate information coding, introduced by R. It may contain up to nine characters: one letter, one sign, and up to seven numerals ($R\ \times\times\times.\times\times\times\times$ inches, $R\ \times\times\times\times.\times\times\times$ millimeters). It is used to control the positions of the Z slide at rapid traverse during positioning mode.

I   The center point coordinate in circular interpolation of the X axis is introduced by I. It may contain up to nine characters in the coding: one letter, one sign, and up to seven numerals ($I\ \times\times\times.\times\times\times\times$ inches, $I\ \times\times\times\times.\times\times\times$ millimeters).

J   The center point coordinate in circular interpolation of the Y axis is introduced by J. This is identical to the I word, only using the J address.

K   The center point coordinate in circular interpolation of the Z axis is introduced by K. This is also identical to the I word, only using the K address.

B   Beta axis coding of the index table, introduced by B. This is a seven-character code: one letter and six numerals ($B\times\times\times\times\times$). It determines the angular position of the index table.

F   Feed rate coding for X, Y, and/or Z axes, introduced by F. It may contain up to five characters in the code: one letter and up to four numerals ($F\times\times\times.\times$ inches/minute, $F\times\times\times\times$ millimeters/minute). It is used for controlling the rate of longitudinal, vertical, and cross travel. The F word is also used in conjunction with the G04 code to dwell the slides. In this

	mode, the format can vary from 0.01 second to 99.99 seconds of dwell.
S	Spindle speed coding for rate of rotation of the cutting tool, introduced by S. It may contain up to five characters in the code: one letter and up to four numerals ($S\times\times\times\times$). This is the actual rpm desired for cutting.
D	Tool trim coding of the tool axis is introduced by D. It may contain up to three characters in the code: one letter and up to two numerals ($D\times\times$). This selects the stored value to be applied to the Z-axis positions for a given operation (optional feature).
T	Tool number coding introduced by T. It is a nine-character code: one letter and eight numerals ($T\times\times\times\times\times\times\times\times$). It determines the next tool to be used.
M	Miscellaneous function coding, introduced by M. It is a three-character code: one letter and two numerals ($M\times\times$). It is used for various discrete machine functions.

PALLET AND PART LOADING AND PROGRAMMING OPTIONS

Part and pallet loading schemes related to stand-alone, cell, and FMS applications vary considerably among users. Various types of fixturing and tombstoning are used to configure and locate prismatic parts for machining center part processing (see Chapter 15).

Use of standard tombstoning, which allows parts to be mounted on a tombstone (Figure 10-15) and machined in one setup, is increasing because of a surge in the popularity of palletized machining. Palletized machining basically permits changing the machine bed. While one pallet of parts is being set up off-line, another is running at the machine. This concept is widely accepted and used, particularly in NC machining center cells and FMS applications.

Parts may be mounted to a standard four-sided tombstone for horizontal machining center processing using three conventional schemes:

1. Single Load Mounting (Figure 10-16a)

One part is loaded on one side of a four-sided tombstone, providing access to only one surface at a time without indexing the *B* axis. This scheme is highly inefficient and does not take full advantage of the tombstone's remaining three sides.

2. Multiload Mounting (Figure 10-16b)

Multiple parts are loaded on two, three, or four sides of a tombstone. This scheme is very efficient and takes full advantage of the tombstone's four sides. The same surface of multiple parts is presented for machining at each *B*-axis index since the tombstone is loaded with parts all mounted identically. Once the lot of parts has been completed for machining similar surfaces, subsequent part settings exposing different surfaces of the same part can be multiloaded on the same tombstone for machining. Or subsequent part settings exposing different surfaces of the same part can be mounted on a different tombstone, pallet, and machine (if available) and processed simultaneously.

3. Progressive Load Mounting (Figure 10-16c)

Progressive loading also takes full advantage of the tombstone's four sides. It involves progressively moving the part through different settings, exposing all sides of the workpiece and removing a completed part at the end of each programmed machining cycle.

Selection of a tombstone pallet or part loading scheme will vary considerably among cell and system users depending on part type, size, and weight, accuracies to be maintained, average lot size, machine and tooling capability and capacity, and other factors. Each cell or system user must examine his own application and circumstances to arrive at the most economical, efficient, and effective loading scheme.

The primary limitations to tombstone pallet and part loading are machine size and weight capacity. Tombstones, in some cases, are higher than the Y-axis travel limits of a horizontal machining center. Consequently, care must be taken not to buy a tombstone whose height dimension exceeds usable Y-axis travel. And caution must be exercised relative to a machine tool's indexable weight limitation and

Figure 10-15 Multiple parts may be mounted to standard tombstones, thus increasing productivity and machine utilization rates. (Courtesy of Remington Arms)

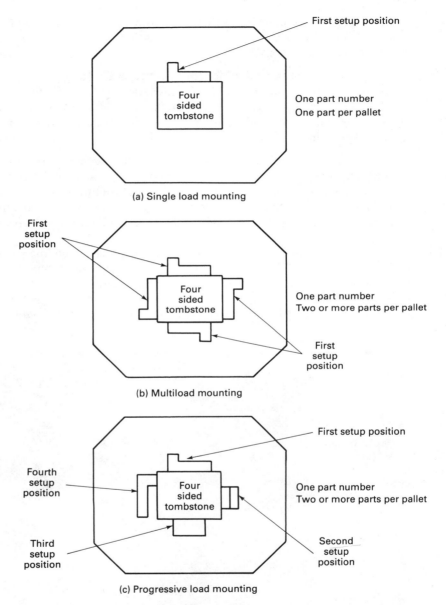

(a) Single load mounting

(b) Multiload mounting

(c) Progressive load mounting

Figure 10-16   Three conventional pallet–part loading schemes are generally used with four-sided tombstones.

an AGV's transport weight capacity. Once multiple parts, fixtures, clamps, nuts, bolts, and washers are bolted to a tombstone, its stand-alone weight can double or triple. Pay attention to the cumulative weight factor when working with tombstones and pallets.

Understanding programming options is important to part processing cycle time and overall machining efficiency, regardless of whether one of the processor languages such as APT or an interactive graphics-based programming package are used. Programming option selection refers to the particular method and approach of cutting tool usage for processing multiple parts in one setup at each machine.

Programming options for processing multiple parts on a four-sided tombstone, for example, can be either tool dominant or part dominant. Tool-dominant programming implies using each cutting tool at every part location requiring that specific tool on all four sides of the tombstone before exchanging that cutting tool for another. Simply stated, indexing the part or tombstone is less time consuming and more efficient than changing the same tool several times during the program.

Part-dominant programming implies using and exchanging all the cutting tools to complete a particular part or tombstone side before indexing to another part or side of the tombstone. Simply stated in this case, exchanging cutting tools, possibly several times on one tombstone side, is less time consuming and more efficient than indexing the part or tombstone and using the same tool in all required locations before exchanging it for another.

Neither tool- nor part-dominant programming is better or more efficient than the other in all cases. In some circumstances, because of the type, variety, and number of cutting tools required for specific parts, both are used within the same program. And, because of the random tooling feature available on modern machining centers, part programmers can call for a tool change whenever desired. However, machining efficiencies may be either gained or lost depending on whether tool or part-dominant programming is used. Critical methods engineering analysis should be employed to study the particular part, setup, and cutting tool application and determine the best techniques for optimal part processing efficiency.

Regardless of how part programming is accomplished for a CNC machine tool, part program prove-out (for turning centers and machining centers) must be completed prior to storing the program for later use. This means that the part program must be run on the machine with a part and tools loaded to see if that specific program produces a part to the dimensions and tolerances required. The part program is made error free and optimized for machining efficiencies prior to release of the program for production use. In general, all part programs are "proved out" prior to bringing a cell or FMS into production. However, in many cases, new parts are added later to a cell or system. As new parts are added, one machine in an FMS is usually taken out of production for program prove-out. Production work is routed to other machines until the program prove-out is completed, errors are corrected, and machining processes are optimized.

AUTOMATED FEATURES AND CAPABILITIES

Three principal developments in the 1960s led to increased acceptance and use of machining centers:

1. The capability of a machine tool to change its own cutting tools on command
2. An indexable work table permitting machining on multiple sides of a workpiece in one clamping (increased versatility of horizontal over vertical machining centers)
3. Calling up interchangeable pallets from an off-line bank for machining center part processing (Williamson's System 24)

Today's trend is to incorporate many diverse functions in a single machining center, such as tilt tables, swivel spindles, and touch-trigger probing. The result is the emergence of the flexible manufacturing cell resident in a single machine: the machining center.

Machining center automated features and capabilities perform a multitude of functions that now automate what was previously performed manually in several separate operations and on a variety of different machine tools. The principal automated machining center features and capabilities are:

1. Torque Control Machining

Sometimes referred to as adaptive control, this feature was developed to speed up or slow down a cutting tool while the tool is engaged in the actual cutting operation. The function of torque control machining is to sense machining conditions (Figure 10-17) and adjust the feeds and speeds to suit the real-time condition. Sensing devices are built into the machine spindle to sense torque, heat, and vibration. These sensing devices provide feedback signals to the MCU, which contains the preprogrammed safe limits. If the preprogrammed safe limits are exceeded, the MCU alters the feeds and speeds.

2. Surface Sensing Probe

Probing is used extensively on machining centers in stand-alone, cell, and FMS applications. Just as in rotational cells and systems, probing is of critical importance to the successful operation of an automated cell or system for prismatic parts. As discussed earlier, the probe is a multidirectional precision electronic switching device that can be held in the tool-storage matrix (Figure 10-18) until called for by the part program. It is then positioned in the machine spindle just as if it were a standard tool. With appropriate CNC-resident software and part programming, machining center probes can:

 ☐ Check for part presence and alignment on single- to multiple-part loaded pallets

Figure 10-17 Torque control machining senses machining conditions and adjusts the feeds and speeds, up or down, to suit the real-time cutting action of each tool. (Courtesy of Maho Machine Tool Corp.)

Figure 10-18 The surface sensing probe can be held in the tool storage matrix until called for by the part program. (Courtesy of Cincinnati Milacron)

□ Calculate the center position of a hole by averaging measured points taken around the hole or boss (Figure 10-19)

□ Compute and store offset data in the fixture offset table

□ Detect stock variations or cored hole shifts and automatically regrid the machine

Probing can improve the machining accuracy by feeding back offsets to fine tune the program in the range of 0.0001 inch or finer. This technique bypasses the need for extremely fine (and costly) drives and position-measuring devices in the machine tool.

3. Automated Tool Delivery

Automated tool delivery to a machining center as part of an automated cell or system offers big gains in productivity and machine utilization rates because machines do not need to be stopped for tool replacement. Generally delivered on a AGV to the rear of the machine and tool matrix (Figure 10-20), new cutting tools can be exchanged with used tools without any interference with the ongoing machining process. Although generally controlled through the added help of a cell controller or minicomputer, the automated tool delivery and exchange capability add considerably to a machining center's overall uptime and performance.

Figure 10-19 Probes can be used for a variety of applications on machining centers, from detecting part presence to calculating the center position of a hole or boss. (Courtesy of Maho Machine Tool Corp.)

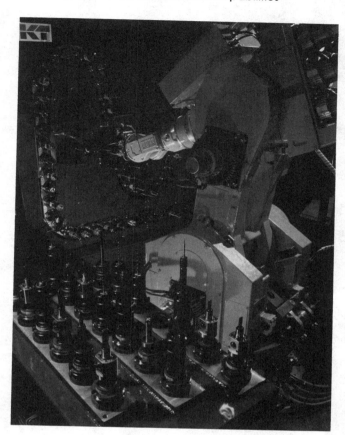

Figure 10-20 Automated tool delivery generally occurs at the rear of the machine and tool matrix and eliminates stopping the machine for tool replacement. (Courtesy of Kearney and Trecker Corp.)

4. Multiple and Angled Spindle Heads

When cycle time can be improved by drilling or tapping several holes at once, multiple spindle or cluster heads may be used. Multiple spindle heads (Figure 10-21), which can be loaded like an ordinary tool, drive a cluster of tools through their internal gearing mechanisms. The head design contains a fixed number of driven spindles, but the location of each spindle relative to the others is determined by the purchaser, who can configure each individual spindle location to suit a repetitive hole pattern. Ninety and forty-five degree angle heads (Figure 10-22) are used on machining centers in highly specialized applications. In most cases, they are used where the investment in an angled spindle head to drill a few holes, for example, may save another complete setup and part handling, just to drill a few difficult to get to holes.

Figure 10-21 Multiple-spindle heads can be loaded like an ordinary tool and drive a cluster of tools to machine a repetitive hole pattern. (Courtesy of Cincinnati Milacron)

Figure 10-22 Ninety and forty-five degree angle heads are used in highly specialized applications. (Courtesy of Cincinnati Milacron)

Figure 10-23   Broken tool detection checks for tool breakage before or after the cutting sequence. (Courtesy of Mazak Corp.)

5.   Broken Tool Detection

This feature, through the NC part program, permits offsetting each tool to a fixed probe to check for tool breakage before beginning the cut sequence (Figure 10-23). If a tool is broken, the machine will automatically replace it with a duplicate stored in the tool matrix. If a duplicate does not exist, a machine-stop condition will occur and operator action is required. Broken tool detection helps to increase machine productivity and utilization and decrease operator involvement and attention.

SUMMARY

1. Machining centers, just like turning centers, are classified as either vertical or horizontal.

2. Both vertical and horizontal machining centers have advantages and disadvantages; the selection of either a vertical or horizontal machining center mainly depends on the part type, size, weight, and application.

3. Horizontal machining centers are generally used in more cell and FMS applications due to the added capability of a rotating index table (*B* axis).

4. Machining center construction has improved to accommodate higher spindle speeds, feeds, and horsepower requirements, along with overall higher utilization rates and increased performance requirements.

5. The primary characteristic demanded of modern machining centers is increased spindle speed.

6. The primary axes of both vertical and horizontal machining centers are X, Y, and Z.

7. Parts may be mounted to a standard four-sided tombstone for horizontal machining center processing using the single-load, multiload, or progressive-load application.

8. The primary limitations to tombstone pallet and part loading are machine size and weight capacity.

9. Programming options for processing multiple parts on a four-sided tombstone can be either tool dominant or part dominant.

10. Modern machining centers incorporate many diverse functions in a single machining center, such as tilt tables, swivel spindles, and touch trigger probing.

11. The principal automated machining center features and capabilities are torque control machining, surface sensing probe, automated tool delivery, multiple and angled spindle heads, and broken tool detection.

REVIEW QUESTIONS

1. _____ machining centers offer easy and accessible pallet shuttle transfer when used in a cell or FMS application.

2. Modern machining center improvements have not yielded much in added safety features or noise-level reduction. True or False?

3. Due to automated tool delivery capabilities, less tool storage capacity is required on modern machining centers. True or False?

4. _____ tool selection is by far the most predominant type used on machining centers.

5. Z-axis movement on a horizontal machining center is:
   (a) Up and down movement of the spindle head
   (b) In and out movement of the machine tool's saddle, table, or spindle carrier
   (c) Up and down movement of the knee or spindle carrier
   (d) None of these

6. Most rotary tables are bidirectional and will index using the shortest path to any of 72, 360, or 720 positions. True or False?

7. The center point coordinate in circular interpolation of the Y axis is introduced by the work address _____ .

8. Palletized machining basically permits changing the _____ .

9. Which type of part and pallet mounting is more efficient and why?

10. The primary limitations to tombstone pallet or part loading are _____ and _____ .

11. Explain the difference between tool-dominant and part-dominant programming.

12. The automated machining center feature that will speed up or slow down a cutting tool based on given machining conditions is known as _____ .

11

Cleaning and Deburring Equipment

OBJECTIVES

After studying this unit you will be able to:

- ☐ Identify wash station types and explain their operation.
- ☐ Name and describe the most common types of automated deburring.
- ☐ Discuss the importance of part cleaning and deburring to automated manufacturing.

INTRODUCTION

Cleaning and deburring equipment, more often than not, are the forgotten members of a cell or system's processing machine group. This is because they are typically considered to be insignificant, after the fact additions to the cell or FMS, performing ho-hum postmachining operations of little value.

Actually, cleaning and deburring equipment do perform ho-hum postmachining operations, but the automated processes add value, save time, and free employees to perform more meaningful work elsewhere. Parts must be cleaned and deburred before they can ever attempt to be accurately inspected, stocked, or assembled. Fixtures and pallets must be cleaned in order to accurately locate succeeding parts.

Although sometimes challenged because simple processes require high-priced equipment, how parts are cleaned and burrs removed are important factors to be considered in the planning of many flexible cells and systems.

WASH-STATION TYPES AND OPERATION DESCRIPTION

Wash stations come in a variety of types, styles, and configurations, but are generally classified as either batch or in-line conveyorized. Batch washers (Figures 11-1 and 11-2) are available to handle workpieces weighing thousands of pounds and as large as a 72-inch cube. Batch wash stations are generally used in low- to mid-volume applications to provide a clean part for downstream inspection, assembly, or further processing.

In-line conveyorized washers (Figures 11-3 on page 223 and 11-4 on page 224) are used for high-volume production where rapid part throughput is a high requirement. With an in-line conveyorized washer, parts are loaded at one end of the system, cleaned as they pass through the machine, and removed at the opposite end. Separate roller conveyors can be added at the load–unload sections for interfacing with a robot or pallet shuttle mechanism. Multiple stages can be added for rinsing, rust prevention, or part blow-dry.

Figure 11-1 Typical batch wash station. (Courtesy of Cincinnati Industrial Machinery)

Figure 11-2   Batch wash stations are used in low- to mid-volume applicatons. (Courtesy of General Dynamics Corp., Fort Worth Division)

Selection of either a batch or in-line conveyorized wash station is a function of:

1. Workpiece type, size, weight, material, and configuration
2. Throughput rate required
3. Material to be removed (chips, cutting oil, tapping compound, and the like)
4. Succeeding operation type (inspection, stocking, assembly, or another machining operation)
5. Method of part loading, unloading, transport, and delivery

Use of either a batch or in-line conveyorized wash station requires spray nozzles to be properly sized, located, and directed to clean exterior as well as recessed and hard to get at interior areas of the workpiece. Adequate volume and pressure are required for complete flushing of chips from the workpiece fixture and pallet. Many high-pressure wash–cleaning stations operating at 400 psi or more are

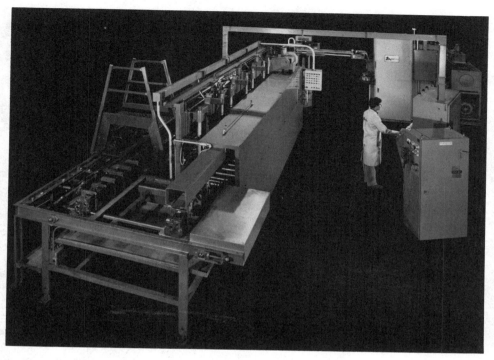

Figure 11-3 In-line conveyorized wash station. (Courtesy of Robert Bosch Corp.—Surf Tran Division)

capable of shearing off encrusted dirt and grease, resulting in a well-cleaned part, fixture, and pallet. However, cleaned workpieces must be checked occasionally to make sure blind holes and recessed areas are clean (these are the most difficult to flush and clean). Consequently, parts should be fixtured and oriented to allow for the most efficient cleaning and draining.

Some heavy-duty batch wash stations are capable of automatically locking the pallet assembly to an internal machine circular rail carriage and rotating the entire assembly around the rail during the cleaning and blow-dry cycle. This allows better access to recessed areas, improved drainage, and increased blow-dry coverage.

Blow-off is one of the most important wash-station options. This reduces drying time of the washed workpiece by blowing off the excess coolant or wash solution, prevents spillover to other machines and other areas of the cell, and helps keep the area clean and neat. Some machines use convector heated air blow-off generated by gas, steam, or electricity in order to speed up the blow-off and part drying cycle and to remove moisture.

In an FMS with a central coolant and chip-removal system, chips and coolant

Figure 11-4 In-line conveyorized wash stations are used in high-volume production applications. (Courtesy of Robert Bosch Corp.—Surf Tran Division)

from the batch wash station flow directly into the flume system trough to be circulated back to the central coolant storage tank. In-line conveyorized wash stations generally have their own individual coolant and cleaning solution storage tanks equipped for chip recovery and coolant or cleaning solution recirculation.

Because a cleaning machine is basically a dirt collector and even though chips are removed, tank clean-out, regardless of type, is very important. Screens and baffles are used to filter metal filings, dirt, sludge, and other contaminants, but these must be periodically cleaned as well. Consequently, tank access and ease of cleaning are principal concerns when purchasing cleaning equipment. Where the volume of dirt removal by the cleaning machine is excessive, continuous clean-out can be provided. A sludge conveyor (Figure 11-5) can be used on any type of wash station to handle any volume of dirt, chips, and the like. Sludge conveyors basically carry the waste material up a slope to be deposited in a sludge container for disposal, while the liquid drains back into the central storage tank.

Wash stations, like the other equipment in a cell or system, receive instructions from the host computer or cell controller to their individual programmable controller. These instructions consist of signals primarily to:

Figure 11-5   Sludge conveyors help keep wash stations clean by removing sludge waste. (Courtesy of Cincinnati Industrial Machinery)

1. Open and close the entry door
2. Position pallet and internal locks before beginning the wash cycle
3. Monitor, control, and feed back tank coolant and cleaning solution levels to the host computer

The wash cycle is then under machine control until completion of the operation. A typical batch wash station operational scenario in an FMS would be:

1. A fixtured pallet is delivered via AGV to the wash station shuttle mechanism.
2. The entire pallet assembly is shuttled to the front of the wash station.
3. Wash station doors open and the pallet assembly moves into the wash zone, where internal locks are activated to lock the pallet to the overhead rail carriage and the doors are shut.
4. The locked fixtured assembly begins its 360-degree overhead rotational path.
5. High-pressure coolant begins to flood the entire pallet assembly through many nozzles, causing chips, dirt, and grease to fall into the flume system.
6. After a timed period and with the assembly still rotating, the coolant flush is shut off and the heated air blow-off cycle begins.

7. The blow-off fans continue for a timed period, and the cycle ends as the carriage returns to its original position.
8. Pallet locks are released, the wash-station doors open, and the fixtured pallet assembly is returned to the shuttle position for disposition to another area.

Some FMS systems do not use any special equipment for cleaning. Machine tools using high-pressure coolant are used to flush the entire part–fixture assembly once machining is completed while still in the machine's work zone. Pressurized air directed through the spindle blows away excess coolant before a pallet change is made. This approach eliminates the extra cost of a separate wash station, part move and queue time, and potential bottlenecks, but offsets chip-making time for part washing and cleaning.

DEBURRING-STATION TYPES AND OPERATION DESCRIPTION

Although great strides have been made in machining over the last several years, one seemingly insignificant process continues to be vital: burr removal. Even the most accurate and finely finished part is rarely usable until deburring has been completed.

In the past, deburring was purely a manual, repetitive, and mundane process that was highly labor intensive. In many instances today, circumstances still require extensive manual effort for part deburring. However, over the last several years, various high-tech methods have emerged or have been perfected to deal with the problem of automating and reducing the manual effort required for burr removal. The four most common types of automated deburring are:

1. mechanical
2. vibratory
3. thermal energy
4. electrochemical

Mechanical deburring involves the use of industrial robots in an automated cell or system. Parts are deburred by a wrist-held wire brush or grinding wheel (Figure 11-6). Rotational power is supplied to the wire brush or grinding wheel while the robot's articulated arm positions the deburring tool in all its proper positions to remove the burrs. The process can be designed to have the robot change deburring tools at a stand-alone tool storage rack to suit the specific part and burr-removal requirements.

In many cases, conventional robots are not ideally suited for burr removal. This is because the robot's articulated arm does not provide enough rigidity and accuracy for heavy burr-removal applications. And many workpieces require different methods of deburring for different parts of the workpiece.

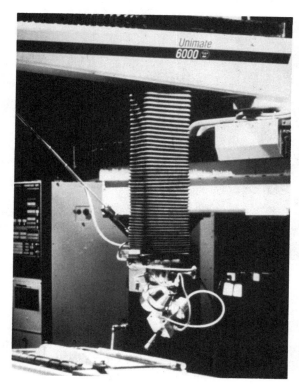

Figure 11-6   Robots are used in many applications for mechanical part deburring. (Courtesy of Westinghouse Automation Division)

Vibratory deburring machines are designed for relatively small rotational or prismatic workpieces. Parts systematically enter a large bowl container (Figure 11-7) filled with ceramic pebbles commonly referred to as media. The size of the ceramic media can vary depending on the type, size, and material of the parts to be deburred. As parts enter the bowl, sometimes via a conveyor, the rapid vibratory back and forth motion agitates the parts in the ceramic media, removing burrs, descaling, and gently polishing the parts. Eccentric weights are mounted on each end of the container support shaft to vibrate the bowl in a controlled but adjustable manner.

Vibratory deburring systems provide agitation flexibility during the cycle: vigorous action for heavy burrs to a gentle action for polishing and micro finishing of delicate parts. Parts can be removed from the vibratory finishing process either manually or by means of a magnetic separator.

Thermal energy deburring uses intense heat to deburr and/or deflash parts. Parts to be processed are sealed in a chamber that is pressurized with a mixture of combustible gas and oxygen that completely envelopes the parts and surrounds burrs and flash, regardless of external, internal, or blind hole location. This gaseous mixture is then ignited by a spark plug, which creates an instant burst of intense heat, and burrs or flash, because of their high ratio of surface to area mass, burst

Figure 11-7   Vibratory deburring machines use ceramic media and agitation to remove burrs. (Courtesy of Automated Finishing)

into flames. Burrs and flash are instantly oxidized and converted to powder in a total floor to floor cycle time of approximately 25 to 30 seconds. Parts can then be cleaned with a solvent. A typical thermal energy deburring station and some example parts can be seen in Figures 11-8 and 11-9 (on page 330).

Thermal energy is a unique and consistent deburring process because it removes undesirable material from all surfaces, even inaccessible internal recesses and intersecting holes. And it eliminates follow-up inspection necessitated by inconsistent hand deburring operations. Thermal energy deburring is effective on a wide range of dissimilar parts of both ferrous and nonferrous material.

Electrochemical deburring machines (Figures 11-10 on page 330 and 11-11 on page 331) can deburr and contour parts through an electrochemical reaction that dissolves metal from a workpiece into an electrolyte solution. Direct current is passed through the electrolyte solution between the electrode tool (the shape of the cavity desired), which has a negative charge, and the workpiece, which has a positive charge. This causes metal to be removed ahead of the electrode tool as the tool is fed into the work. Chemical reaction caused by the direct current in the electrolyte dissolves the metal from the workpiece. Burr removal and chamfering–contouring are adjustable such that high current produces a high rate of metal removal and low current produces a low rate of metal removal.

Although electrochemical deburring and contouring are slow processes, they have several advantages. The tool (electrode) never touches the part, so no tool wear occurs. No heat is created during the process; therefore, thermal or mechanical stress cannot distort the part. And electrochemical deburring is applicable across a wide range of material types and hardness variations.

Figure 11-8   Thermal energy deburring machine. (Courtesy of Robert Bosch Corp.—Surf Tran Division)

IMPORTANCE TO AUTOMATED MANUFACTURING

Planning the design and implementation of any cell or system requires careful, systematic study of the many factors that affect final results. Among these key factors is the method for removal of burrs, chips, dirt, grease, tapping compound, and coolant from parts, fixtures, and pallets.

The burr-removal and cleaning process is of vital importance as chip- and burr-free part surfaces are a requirement for accurate and trouble-free fixturing, inspection, and assembly. It is a necessary part of the part manufacturing cycle normally thought of strictly in terms of manual effort. If parts cannot be deburred and cleaned in an automated cell or system, they must be routed external to the cell

Figure 11-9 Thermal energy deburring uses intense heat and an ignited gaseous mixture to deburr and deflash parts. (Courtesy of Robert Bosch Corp.—Surf Tran Division)

Figure 11-10 Electrochemical deburring machine. (Courtesy of Robert Bosch Corp.—Surf Tran Division)

Figure 11-11 **Electrochemical deburring uses a chemical reaction caused by the direct current in the electrolyte, which dissolves metal from the workpiece. (Courtesy of Robert Bosch Corp.—Surf Tran Division)**

to be manually cleaned and deburred and then routed back into the cell or to another separate area for inspection. Such activity causes excessive part move and queue time, along with adding humdrum labor-intensive work for the employees who must perform these functions. Although not always possible, the objective should be to eliminate as much manual piece handling and labor effort as possible.

As we have already seen, sophisticated equipment has been developed and is being applied to automate the deburring and cleaning process and eliminate the time-consuming human drudgery associated with these tasks. However, in many instances both part cleaning and deburring are difficult to automate and also interface with the material-handling system. Consequently, they sometimes too easily fall into the categories of being "impractical," "too difficult," or "the cost isn't worth the benefits."

Taking the time to closely analyze and study the value of adding part cleaning and deburring into an automated cell or system can bring about a change of mind, because it can have a considerable effect on quality, productivity, and delivery. It is important to consider adding these processes to an automated cell or system because they can:

1. Eliminate the manual move, queue, labor, and piece-handling time
2. Improve part flow and throughput
3. Provide a cleaner and safer work environment
4. Reduce potential part damage as a result of the extra manual part handling
5. Free personnel for more meaningful tasks
6. Add more control to the total part manufacturing process

Of the two processes, cleaning and deburring, cleaning is more flexible and generally easier to add to a cell or system than deburring. Depending on part characteristics and other factors, it may not always be cheaper.

Deburring has limited flexibility of operation, as we have already seen. Different types of deburring may be required for different parts of similar workpieces. If workpiece requirements change, the method and type of deburring may have to change.

Wash stations, on the other hand, can accommodate a variety of different parts, as long as the parts can fit within the required size limitations. And batch wash stations must be able to accommodate the height and weight of tombstone fixtures. Consequently, how large a part and tombstone fixture can be accommodated by a particular wash station is an important factor to be considered in purchasing.

SUMMARY

1. Sophisticated processes and equipment have been developed to automate part cleaning and deburring and eliminate the time-consuming human drudgery associated with these tasks.

2. Cleaning stations are automated high-tech washing machines that use high-pressure coolant or cleaning solvent to remove dirt, grease, and chips from the part, fixture, and pallet.

3. Wash stations are classified as either batch or in-line conveyorized.

4. Batch wash stations are generally used in low- to mid-volume applications, while in-line conveyorized washers are used for high-volume production.

5. Blow-off is one of the most important wash-station options because it blows off excess coolant or cleaning solvent and reduces drying time.

6. Machine tools with high-pressure coolant flush and pressurized air blow-off capabilities are used in some FMS systems to eliminate the extra cost of a separate wash station.

7. The most common types of automated deburring are mechanical, vibratory, thermal energy, and electrochemical.

8. In many instances today, circumstances still require extensive manual effort for part deburring.

9. The burr-removal and cleaning process is of vital importance as clean and burr-free part surfaces are a requirement for accurate and trouble-free fixturing, inspection, and assembly.

REVIEW QUESTIONS

1. The type of wash station that is used where rapid part throughput is a high requirement is the _____ .

2. The method of part transport is not a factor to be considered when selecting a batch wash station. True or False?

3. All wash stations equipped with air blowoff use heated air. True or False?

4. Batch wash stations have their own coolant and cleaning solution storage tanks. True or False?

5. _____ are used with some wash stations to carry away sludge waste and allow the cleaning liquid to drain back into the storage tank.

6. Briefly describe the batch wash-station operating scenario.

7. Some modern means of deburring enable burrs to be removed from inaccessible internal recesses and intersecting holes. True or False?

8. Name the most common types of automated deburring.

9. Thermal energy deburring:
  (a) Requires the use of an electrolyte solution
  (b) Is not ideally suited for burr removal from internal recesses
  (c) Uses intense heat and an ignited gaseous mixture to deburr and/or deflash parts
  (d) Requires agitation in ceramic media to remove burrs

10. Explain the importance of adding automated cleaning and deburring to a flexible cell or system.

12

Coordinate
Measuring Machines

OBJECTIVES

After studying this unit, you will be able to:

☐ Identify CMM types and explain their general function

☐ Describe the operational cycle of a CMM

☐ Discuss the importance of CMMs to flexible cells and systems

INTRODUCTION

In a span of less than 10 years, coordinate measuring machines have evolved from simplistic manual machines that required hours for an inspector to check a part, to sophisticated high-tech inspection centers that can automatically inspect a part in minutes. However, inspecting parts faster is only part of the value that CMMs can provide. Coupled with processing machines in an automated cell or system, CMMs provide many unique capabilities, such as automatically comparing inspection results with preestablished tolerance bands and dynamically linking inspection data to machine tool offsets to compensate for tool wear without operator intervention.

Although sometimes considered to be nonvalue-added components of a cell or FMS-like wash stations, CMMs do add value to the part manufacturing process. And, when statistical process control is applied to the inspection process, these data can be used to virtually eliminate bad parts.

The acceptance and use of CMMs in both stand-alone and automated system applications have been rapid. Their growth in the active production processes will continue as both the machine technology and software mature even further in the years ahead.

TYPES, CONSTRUCTION, AND GENERAL FUNCTION

Coordinate measuring machines, commonly referred to as CMMs, come in a variety of types, sizes, and models. Just like CNC machine tools, CMMs are classified as either vertical or horizontal. Vertical CMMs (Figures 12-1 and 12-2) are sometimes referred to as bridge models, while horizontal CMMs (Figures 12-3 on page 237 and 12-4 on page 238) are occasionally called cross-bar or cantilever models. Both types have various manufacturers and range in size from small table-top models (Figure 12-5 on page 238) to the very large and expensive floor-

Figure 12-1   Vertical CMM. (Courtesy of Sheffield Measurement Division)

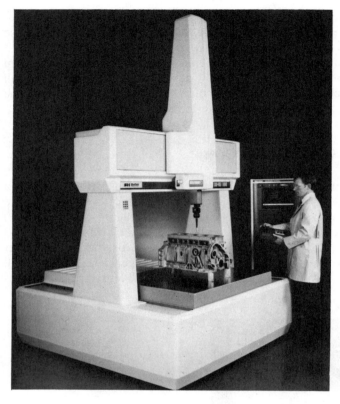

Figure 12-2   Vertical CMMs can be referred to as bridge-type inspection machines. (Courtesy of Sheffield Measurement Division)

mounted kind (Figure 12-6 on page 239). CMMs are available with various computer peripherals and offer a variety of improved software packages, making systems integration of this equipment more practical.

CMM construction in the past was compromised by the inherent limitations of the materials used. However, the materials used for CMM construction have been improved considerably in recent years, greatly improving machine accuracy, precision, and repeatability. The measuring table and all the guideways, for example, are constructed of high-quality granite (Figure 12-7). Increased use of ceramics is also gaining wide acceptance among manufacturers to enhance thermal stability and reduce vibration, thereby improving accuracy. The advantages of using granite and ceramics over conventional materials for inspection equipment are:

1. Greater thermal and dimensional stability
2. Improved bearing surfaces
3. Increased strength, rigidity, and structural integrity
4. Less manufacturing downtime and higher productivity
5. Wear-free operation
6. Longer periods between calibration

Figure 12-3 Horizontal CMM. (Courtesy of Cincinnati Milacron)

Floor preparation for coordinate measuring machines is also very important. Solid reinforced concrete foundations are required for vibration dampening. In some cases, these concrete foundations, depending on inspection equipment type and size, are several feet thick. Many CMMs, however, have self-leveling, pneumatic antivibration systems that effectively reduce mechanical vibration and shocks. In some cases this will eliminate the need for a separate and special foundation.

CMMs also require environmental control. Measured dimensions can only be as accurate and reliable as the stability of their surrounding environment. Consequently, some coordinate measuring machines must be in an environmentally controlled room or the machine itself must be completely enclosed (Figure 12-8 on page 240). Maintaining inspection equipment in an environmentally controlled room or enclosure helps keep the machine at a constant temperature for measurement accuracy and repeatability, while preventing contaminants from affecting exposed surfaces and critical components. However, some CMMs are designed to be able to cope with shop floor conditions. Vendors are increasingly incorporating

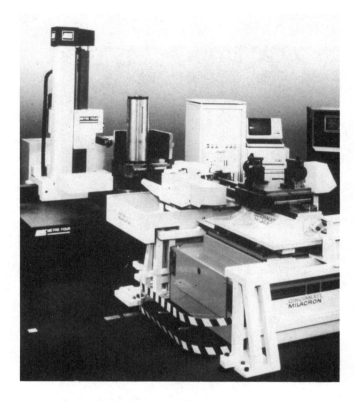

Figure 12-4 Horizontal CMMs are sometimes called cross-bar or cantilever models. (Courtesy of Remington Arms)

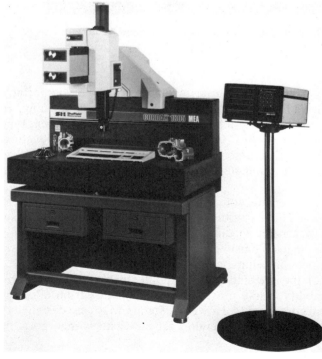

Figure 12-5 Small table-top CMM. (Courtesy of Sheffield Measurement Division)

Figure 12-6 Some CMMs are very large and expensive floor-mounted models. (Courtesy of Cincinnati Milacron)

Figure 12-7 The measuring table and guideways on many CMMs are constructed of high-quality granite or ceramics. (Courtesy of Sheffield Measurement Division)

**Figure 12-8   Some CMMs must be
in an environmentally controlled
room or in a self-contained
enclosure. (Courtesy of Sheffield
Measurement Division)**

temperature- and humidity-compensating capabilities into their machines, as well as contamination protection.

CMM inspection programs are prepared and proved out in advance of actual use just like the CNC part programs that machine the parts. Axes movements in X, Y, and Z are similar to other vertical and horizontal CNC equipment. During the automated workpiece inspection process, part dimensions are recorded with the help of the probe (Figure 12-9). Probes for coordinate measuring machines come in a variety of makes and models to suit the specific requirements of the inspection application (Figures 12-10 on page 242 and 12-11 on page 242). The CMM compares result measurements with the previously input manufacturing tolerances allowed for each dimension and conveys this information to either the host or CMM local computer.

Typically, the CMM computer is interfaced with the FMS host computer for uploading and downloading inspection data and CMM part programs. But the CMM computer plays an important role in the operation of a coordinate measuring machine. In general, its primary functions include:

Figure 12-9   Part dimensions are recorded on CMMs with the aid of the touch-trigger probe. (Courtesy of Cincinnati Milacron)

1. Control of graphical display and hard-copy output of measured data
2. Storage and retrieval of measured data
3. Determination of dimensional deviations
4. Transfer of dimensional data files to the host computer
5. Generation of SPC data
6. Storage of machine calibration data
7. Measurement comparison with test data parameters
8. Operation of the CMM through manual or part program control

OPERATIONAL CYCLE DESCRIPTION

The parts to be inspected by a coordinate measuring machine require prepro-grammed inspection programs for each different workpiece. These typically reside

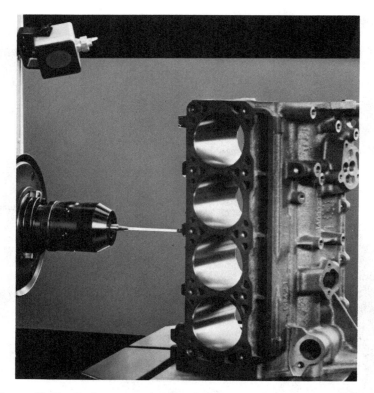

Figure 12-10 Probes come in a variety of makes and models to suit the specific part requirements. (Courtesy of Cincinnati Milacron)

Figure 12-11 Probes are unhampered by complex part geometry and can access a wide variety of internal part details requiring inspection. (Courtesy of Sheffield Measurement Division)

at the host computer level in an FMS. Inspection programs are downloaded on demand to the CMM computer.

Parts are delivered to the inspection station in an FMS after completion of the wash cycle via an AGV. The docking procedure at the inspection station is controlled and monitored by a programmable logic controller (PLC). The PLC ensures that the correct docking procedure has been followed and reports to the host computer that docking has been completed.

Pallet and part identification are verified, and the proper CMM inspection programs are automatically downloaded to the CMM computer. Inspection programs are sometimes downloaded in advance of pallet arrival and are deleted upon completion of the inspection routine. Inspection results are automatically captured and archived via workpiece serial numbers residing on the FMS host computer.

Inspection data from the CMM are automatically compared with pre-established tolerance bands in the CMM part program. Typical inspected features include diameters, hole depths and positions, flatness, and depth of machined area. If a single workpiece on a fixture is rejected by a CMM, that pallet is automatically routed to a material review station. An operator can then visually review the part and the inspection results and dispatch the rejected workpiece pallet to make a second pass through the CMM. If the part is rejected a second time, the rejection is recorded in the associated workpiece history file on the host computer, and the pallet continues its predetermined routing.

In some cases the CNC machine tool that produced the workpiece is placed on hold. Messages are issued to the operating personnel to take appropriate action, such as inspecting tools and fixtures to identify the problem cause. If the workpiece (or several different workpieces on a pallet) passes inspection, the pallet is returned to the CMM's shuttle position for AGV transporting to the queuing carrousel to await part removal or refixture for a subsequent machining operation.

Motorized probe heads are extremely accurate and permit uninterrupted inspection of complex components, as seen in Figure 12-12. Probing contacts rendered doubtful by dirt or chips, for example, are recognized and automatically repeated; missing holes or bores are recorded. In the case of collisions with the part, the probe head returns to its reference position via a number of different safety paths or positions. As a result, specific dimensional checks are either skipped over or the measurement is continued on the next part. Deviations can be plotted to provide both graphical (SPC) and analytical inspection results.

In some cases, on-machine final inspection is performed following completion of machinery, high-pressure coolant flush, and blow-dry. On-machine inspection (Figure 12-13) is accomplished with special measuring peripherals, such as a variety of probes to suit the inspection application and an integrated personal computer with the appropriate quality and statistical process control software. Inspection programs can be downloaded from the FMS host computer to the integrated personal computer, and the results archived or printed as needed.

The advantages of on-machine inspection over individual CMM inspection are:

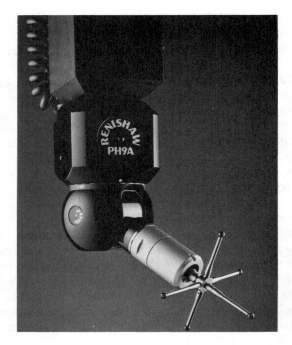

Figure 12-12   **Motorized probe
heads permit internal inspection of
complex components without
complicated probe–stylus
configurations. (Courtesy of
Renishaw, Inc.)**

Figure 12-13   **On-machine
inspection is used in some cells and
systems to avoid the purchase of an
expensive CMM. (Courtesy of
Kearney and Trecker Corp.)**

1. Elimination of an expensive CMM; however, chip-making time is sacrificed for on-machine inspection time
2. Automatic machine tool compensation of dimensional deviations from acceptable tolerances
3. Elimination of part move and queue time to and from the CMM
4. Automatic dimensional checks and compensation that may be necessary due to tool wear or part warpage

IMPORTANCE TO FLEXIBLE CELLS AND SYSTEMS

In the past, traditional inspection of complicated parts was extremely slow, highly prone to error, and difficult to track. It was also expensive. Production equipment sat idle waiting for first-piece inspection, which wasted precious chip-making time and interrupted part throughput.

With the introduction of computer-controlled CMMs, inspection routines that used to be performed manually and take hours to execute are now done swiftly and automatically in minutes, and with greater accuracy and precision. The increased capability, accuracy, and performance have enabled coordinate measuring machines to be successfully integrated into flexible cells and systems as an integral and vital system component (Figure 12-14).

CMMs are important to automated cells and systems not merely to indicate whether the manufacturing process made a good or bad part, but to control the overall process so that it makes only quality parts. This is a radical change from the past, when CMMs were traditionally used as expensive go/no-go gages.

Modern manufacturing demands require parts to be inspected much quicker now. Errors must be found and corrected as quickly as possible and parts need to be inspected as close to the point of actual machining as possible. This means that CMMs must be able to perform accurately and consistently under adverse shop conditions, and parts cannot idly sit around waiting to temperature-stabilize before inspection. Consequently, inspection and process control are integral to the cell or system and manufacturing process, as opposed to being separate or stand-alone as in the past.

Besides being capable of automatically compensating for environmental conditions, CMMs must be faster and more flexible. The variety of workpiece types and sizes that CMMs is required to inspect and the rates at which they must be inspected continue to increase. And CMMs must be designed to be accessible from multiple sides for part loading and unloading, making them easy to integrate into automated material flow systems.

Probe changers are also an important part of a CMM's flexibility and overall importance to automated manufacturing. Many probes are the motorized multipositioning type that can be programmatically adjusted to suit the specific part feature being inspected or interchanged among a variety of probe heads to accommodate part feature differences (Figure 12-15). Probes are unhampered by complex part geometry or surface finish and can access a wide variety of internal part details requiring inspection.

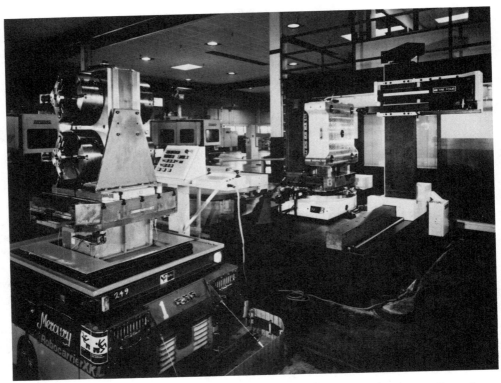

Figure 12-14   CMMs have become an internal component of flexible cells and systems due to their increased capability, accuracy, and performance. (Courtesy of Cincinnati Milacron)

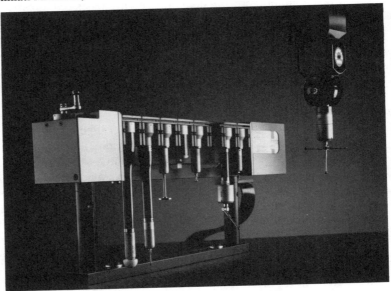

Figure 12-15   Probe head exchange capabilities ensure fast probing to probing cycles to accommodate various part feature differences. (Courtesy of Renishaw, Inc.)

But probing is contact inspection, and the need to physically move the part or the probe to perform measurements imposes significant limitations on data collection rates. As a result, noncontact inspection systems utilizing video and laser inspection capabilities are in use and are being perfected and improved to provide additional speed and flexibility to the overall process control and inspection function.

SUMMARY

1. CMMs are classified as either vertical or horizontal and range in size from small table-top models to very large and expensive floor-mounted types.

2. Many CMMs are constructed of high-quality granite and ceramics that offer several advantages over conventional materials used for inspection equipment.

3. Some CMMs require enclosures for environmental control, while others have self-incorporated temperature- and humidity-compensating capabilities, as well as contamination protection.

4. The CMM's primary function is to compare result measurements with the previously input manufacturing tolerances allowed for each dimension and convey this information to either the host or CMM computer.

5. The CMM computer is interfaced with the FMS host computer for uploading and downloading inspection data and CMM part programs.

6. Docking procedures at the inspection station are controlled and monitored by a programmable logic controller (PLC).

7. In some cases, on-machine inspection is performed through the use of special measuring peripherals, which can eliminate an expensive CMM, but sacrifices chip-making time for on-machine inspection time.

8. Probes come in a variety of types and sizes for automated inspection, many of which are motorized and can be interchanged among a variety of probe heads to accommodate various part feature differences.

9. The increased capability, accuracy, precision, and performance of CMMs have enabled them to be successfully integrated into flexible cells and systems.

REVIEW QUESTIONS

1. Horizontal CMMs are occasionally called _____ or _____ models.

2. Which of the following are advantages of using granite and ceramics over conventional materials for inspection equipment:

  (a) Greater thermal and dimensional stability
  (b) Longer periods between calibration
  (c) Wear-free operation
  (d) Increased strength and rigidity
  (e) Only (a), (b), and (d)
  (f) All of the above

3. Solid reinforced concrete foundations are required for vibration dampening for all CMMs. True or False?

4. Name five primary functions of the CMM computer.

5. Inspection programs are downloaded on demand to the CMM computer. True or False?

6. The CMM computer ensures that the correct docking procedure has been followed and reports to the host computer that docking has been completed. True or False?

7. Inspection results are automatically captured and archived via workpiece serial numbers residing on the _____ computer.

8. Describe what action occurs if a workpiece is inspected and rejected by the CMM.

9. Name three advantages of on-machine inspection over individual CMM inspection.

10. Explain the importance of CMMs to flexible cells and systems.

Part IV

System Support Equipment

As we have already studied, it takes many elements functioning together to make a cell or system operate successfully. Some of these elements are support oriented and do not receive as much attention as the more obvious and value-added equipment, like machine tools and coordinate measuring machines.

System support equipment consists mainly of material-handling equipment, robots, queuing carrousels, chip conveyors and compactors, cutting tools, fixtures, and other related accessories. Some of this equipment tends to be taken for granted in many cases— until something goes wrong. One unavailable cutting tool or defective fixture, for example, can prevent a machine tool from completing an urgently needed part. Then it becomes the most important element in the system.

The importance of support equipment to flexible cells and systems increases in value as the size and complexity of the system increases. Without the attention paid to the necessary support equipment, the value-added processing equipment cannot add much value. Consequently, system support equipment needs to receive its equal share of planned attention— and before, not after, the system is installed.

13

Automated Material Movement and Storage Systems

OBJECTIVES

After studying this unit, you will be able to:

□ Identify primary AGV types and functions in a cell or FMS.

□ Explain how robots are classified and used in automated systems.

□ Describe what an ASRS does and why it is important to a company's automation efforts.

□ Name the primary types of conveyor systems.

□ Discuss the importance of parts queuing, chip disposal, and coolant recovery to an automated system's success.

INTRODUCTION

Because of the waste associated with work in process and finished goods inventory, automated movement, storage, and retrieval of material continue to gain wide attention and acceptance. Material movement and storage, with respect to cells and systems, cover more than the traditional workpiece flow and movement; they also include tool, fixture, and pallet movement and storage to and from the processing stations and queue areas along with chip collection and removal. Such widespread demand places an increasing burden on any material movement and storage system and shifts the focus of attention from actual part manufacturing to the handling, control, movement, and storage of inventory and system accessories.

In smaller cells and systems, robots, interfaced with machine tools, continue to gain wider acceptance and use. In many cases a robot interfaced with one or more machine tools is a company's first attempt at a cell. Robots are application dependent and when applied to repetitive, limited flexibility cells are very stable and reliable material handlers. And they can be reprogrammed for other tasks or interfaced with other equipment as requirements change.

Material-handling, movement, and storage systems continue to be developed and improved as increasing attention is focused on the timely and reliable delivery, storage, and retrieval of all types of material and system support equipment. Such emphasis will continue to proliferate in the years ahead as automated material-handling systems increase their value to flexible manufacturing cells and systems.

AUTOMATED GUIDED VEHICLES

Automated guided vehicles (AGVs), as they are commonly referred to, were ironically invented in the United States, but were first used in Europe after World War II as "driverless tractors." AGVs, as we know them today, were developed in the United States in the mid 1950s and are defined by the Materials Handling Institute as battery-powered driverless vehicles that can be programmed for path selection and positioning and are equipped to follow a changeable or expandable guidepath.

In the 1970s, computers affected AGV technology and vehicles became bidirectional, moving material between assembly lines, shipping docks, and processing stations. The computer, communicating via FM radio signals, gave AGVs the ability to travel on both closed and multiple loop paths and also handled traffic control and the queuing of multiple-vehicle systems. This onboard microprocessor and "land-based" AGV computer allowed for material tracking as well.

AGVs come in a variety of types and sizes, as seen in Figures 13-1, 13-2, and 13-3, and can be used in applications and environments wherever material is moved. AGV general types include:

- Towing
- Pallet trucks
- Unit load
- Fork trucks
- Assembly vehicles

They typically can carry from 35 to 120,000 pounds at speeds ranging from 40 to 200 feet per minute. AGVs are becoming increasingly complex and sophisticated devices. Many are guided by onboard computers, and some have vocabularies up to 4000 words and can be programmed in any language.

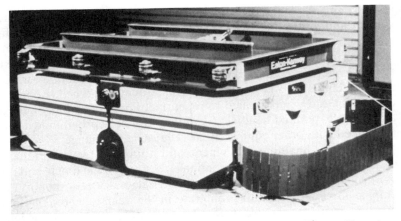

Figure 13-1   AGVs come in a variety of types and sizes. Pictured here is a unit load AGV. (Courtesy of Remington Arms)

Figure 13-2   Fork-lift automated guided vehicle. (Courtesy of Interlake Material Handling Division)

Figure 13-3   Fork-lift AGV removing a chip bin. (Courtesy of Interlake Material Handling Division)

The majority of AGVs in use are battery-powered, wire-guided vehicles that follow energized wire embedded in the floor. Saw cuts are made in the floor about one-half-inch deep based on a predetermined guidepath route. Wire is laid in the saw cut and then epoxied over to form a smooth, unbroken surface for sweeping and maintenance (Figure 13-4). Guidepath routes can conform to virtually any type of shop or warehouse layout but are generally classified as in-line, branched, loop or webbed, as seen in Figure 13-5 on page 255. Carts are propelled by electric motors that are powered by industrial-grade, lead-acid storage batteries mounted in the AGV. They typically have a normal charged cycle life of around 20 hours. Then the cart must be routed to the AGV maintenance area for battery recharge or replacement. Battery replacement must be done manually, but recharging can be done either manually or automatically. Some carts can be programmed and routed to plug themselves in for recharging when battery power becomes low.

Other means of guidance for AGVs range from the earlier and primitive tow-line vehicles (Figure 13-6 on page 256) to the more upscale and modern optical, laser and "teach-type" vehicles (Figures 13-7 on page 256 and 13-8 on page 257). The table in Figure 13-9 on page 258 lists AGVs by their method of guidance.

These modern types of AGV guidance systems are available and gaining in popularity and acceptance because of the groundswell of technological push to get AGVs off wire. However, wire-guided AGV systems are proven, acceptable, and reliable in the United States, Europe, and other countries. And they are very much a part of many different types of businesses that move material.

Figure 13-4 Wire-guided AGVs at a battery recharge station. (Courtesy of Remington Arms)

Automated guided vehicles can be used in a variety of applications ranging from dirty manufacturing shops to clean-room assembly environments. They offer several advantages over conventional material-handling systems, including:

1. Dispatching, tracking, and monitoring under real-time computer control
2. Better resource utilization
3. Increased control over material flow and movement
4. Reduced product damage and less material movement noise
5. Routing consistency but flexibility
6. Operational reliability in hazardous and special environments
7. Ability to interface with various peripheral systems, such as machine tools, robots, and conveyor systems
8. Increased throughput because of dependable on-time delivery
9. Full electrical power systems and design modularity for ease of problem diagnosis and maintenance
10. High location and positioning accuracy
11. Improved cost savings through reductions in floor space, work in process, and direct labor

Many safety features are available on AGVs. Some of the more common include:

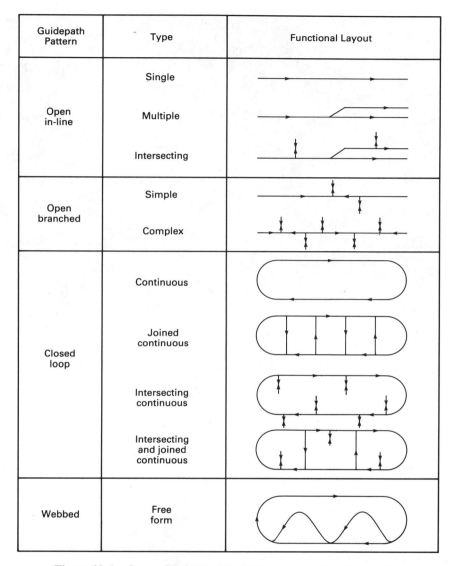

Figure 13-5   General AGV guidepath types for cells and systems.

1. Yellow caution beacons mounted on the front and back of the cart that flash when the cart gets ready to move and that continue throughout the move
2. Audible, multiple-pitch and adjustable warning signals
3. Safety brakes on each drive wheel that automatically engage when the power is off
4. Emergency stop buttons on each side of the vehicle
5. Impact-sensing safety bumpers on each end of the vehicle that stop the cart when minimal contact and bumper compression are made

Figure 13-6   Some of the earlier AGVs were towline vehicles. (Courtesy of Kearney and Trecker Corp.)

Figure 13-7   Optical guidance system AGV. (Courtesy of Eaton-Kenway, Inc.)

6. Operator-controllable, hand-held, plug-in control unit that can direct the cart to perform all functions and move off the wire-guided path
7. Battery low-power message display on the system manager's computer console display screen
8. Enclosures and chip and coolant guards that protect electronic circuits from cutting fluid, chips, and oil contamination
9. Wipers and brushes installed on the underside of the cart that keep the guidepath free of dirt and chips

In an FMS, the AGVs and their centralized computer function as a dispatcher and a traffic manager in a real-time subordinate role to the FMS host computer. The traffic control system must include the capability for automatically tracking the current status of all payloads active in the system. This includes carts, pallet shuttles, queuing carrousels, tool and fixture build and teardown stations, and processing stations. To eliminate AGV congestion and collisions, the guidepath is generally divided into zones with predetermined stop points. Sometimes only one vehicle is permitted in a zone at any one time.

In planning the system, it is necessary to determine the number of carts needed to keep machines supplied with parts without the machines waiting on the material-handling system. Also, enough batteries and recharge stations must be available to provide for daily battery changing and charging on a rotational basis.

**Figure 13-8   Some AGVs are programmable for guidance and movement.
(Courtesy of Interlake Material Handling Division)**

Guidance Type	Description
Tow Line	Primitive mechanical "street car" type of apparatus; chain/cable powered embedded in floor
Wire Guided	Vehicle's antenna senses and follows an energized wire embedded in the floor
Inertial Guidance	Uses an on board microprocessor to steer vehicle on a preprogrammed path; uses sonar sensors for obstacle detection and gyroscopes for directional change
InfraRed	Infrared light is transmitted and reflected from reflectors in the roof of a facility; radarlike detectors relay signals to computer and calculations and measurements taken to determine position and direction of travel
Laser	Laser scans wallmounted, bar-coded reflectors; Through known distances and measurement of the distance the vehicle's front wheel has traveled, the AGV can be accurately maneuvered and located
Optical	Photosensors read and track colorless fluorescent particles painted or taped on concrete, tile, or carpeted floor; often the desired choice of AGV system where maintaining a clean environment is a requirement
Teach Type	Learns guidepath by "walking through"; as programmed vehicle is moved along the desired route, it actually learns the new path and informs the host computer what it has learned; the host computer informs other AGVs of the new path.

Figure 13-9   Several different AGV guidance systems are available; selection is mainly dependent upon application, environment, and need.

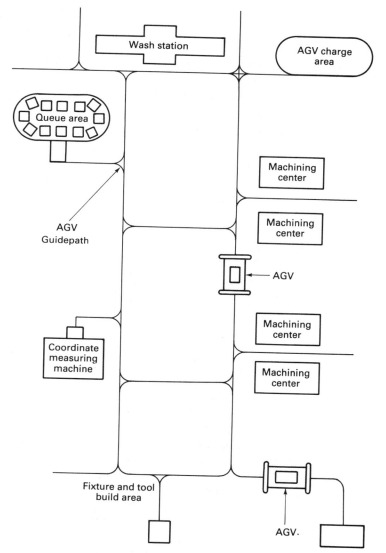

Figure 13-10   AGV guidepaths are a connected series of loops and in-line segments and are laid out as part of the FMS master plan.

An AGV guidepath must be laid out as part of the FMS master plan. Guidepaths are typically a connected series of loops and in-line segments joining the loops, as seen in Figure 13-10. Guidepaths must be wide enough for the AGVs to work their way through the system without any interference between the carts, their payload, and other equipment. Also, guidepaths must be open enough to allow a cart to be sidetracked off-line or bypassed if necessary without wasting overall floor space.

AGVs perform a variety of functions in flexible cells and systems, consisting of:

1. Transporting parts, tools, and fixtures to and from processing, queuing, and build stations
2. Delivering raw material to the cell or system
3. Transporting finished parts from the system to assembly areas
4. Delivering parts, tools, and fixtures to and from an automatic storage and retrieval system (ASRS)
5. Transporting chip containers
6. Automatically raising and lowering pallets to registration positions on processing and queuing station shuttle mechanisms for loading and unloading

AGV technology continues to evolve, driven by continued advances in sophisticated electronics, advanced sensors, and high-tech microprocessors. AGVs are clearly affecting businesses wherever material needs to be moved, and their use will continue to expand in the years ahead.

ROBOTS

The industrial robot is one of the most important developments in the history of automation technology. Although the first articulated arm was developed and used in the 1950s, the industrial robot, as we know it today, did not see wide-scale applications until the development of the microprocessor.

Many definitions are used for the term "robot" and many lay people have their own image of what a robot is and what it can do. In fact, there are those who say that an NC machine is just a single-purpose robot and those that say a robot is just a multifunctional NC machine. However, the definition developed by the Robot Institute of America (RIA) is: "A reprogrammable multifunctional manipulator designed to move material, parts, tools, or specialized devices through variable programmed motions for the performance of a variety of tasks."

Modern robots are much like the original numerically controlled machine tools. Some are limited to simple positioning functions. Some can operate throughout a continuous path, and some can maintain a proper tool orientation regardless of controlled path. Understanding the differences can be critical in making the best selection for the application required.

Robots, like machine tools, are available in a variety of types, styles, and sizes, as seen in Figures 13-11, 13-12, and 13-13. Generally, they are described as either nonservo or servo. Nonservo robots have no servo drives and have limited control over speeding up or slowing down movement. They are very simple, inexpensive robots with limited capability. Nonservo robots have two to four axes of movement, generally with one of the axes rotary, and they move one axis at a time. They operate either by compressed air or hydraulic actuation. Typically, they pick up an object and move it from point A to point B and are controlled by

Figure 13-11 Robots are available in a variety of types and styles. Pictured here is a floor-mounted pedestal robot. (Courtesy of GMFanuc Robotics Corp.)

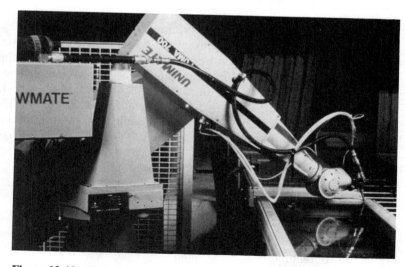

Figure 13-12 Robots are used in a variety of pick and place, welding, and other applications. (Courtesy of Westinghouse Automation Division)

**Figure 13-13   Gantry robots are ideally suited for part loading and un-
loading in machine cells because they minimize floor space. (Courtesy of
Cincinnati Milacron)**

moving against a limited number of hard stops, trip points, or similar devices along
each axis.

    Servo-driven robots are determined by their electric or hydraulic servo
drives. Their capability of being programmed makes them very flexible and open to
a wide variety of applications. Servo-driven models range from simple pick-and-
place units to those with multiple capabilities. Servo-drive robots generally are the
teach-and-lead-through type with the hand-held pendant (Figure 13-14), or they
may be programmed off-line much like NC machines. They are the most commonly
used robots because they are reprogrammable and multifunctional. No industry standards exist for classifying robots, but industrial robots, are generally classified by:

1. Arm configuration and reach
2. Power sources and programmable speed
3. Load capacity
4. Application capabilities
5. Control technique and intelligence

    Also, methods of mounting can vary depending on the application. Although

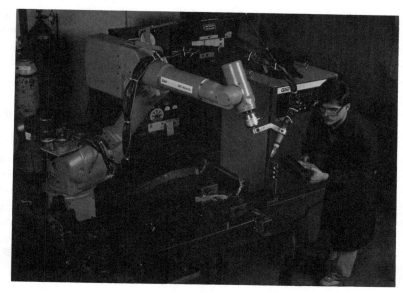

Figure 13-14   Many servo-drive robots are the teach and lead through type with hand-held pendant. (Courtesy of GMFanuc Robotics Corp.)

some robots are movable and can be mounted on a cart that can traverse back and forth, most are fixed. Fixed-mounting types consist of:

1. Floor mounted
2. Machine mounted
3. Wall mounted
4. Gantry

The benefits of using robots include:

1. Increased productivity
2. Improved product quality
3. Application consistency
4. Increased operating hours
5. Reduced scrap and rework costs
6. Direct labor savings
7. Improved worker safety and job enrichment

Although robots and their controls have the normal built-in safety features, such as emergency stop buttons and overload sensors, their repetitive capabilities can create hazardous circumstances. Automation has a tendency to lull individuals into a state of complacency. Consequently, workers could knowingly or unknowingly walk into an automated cell or other robot application and be seriously injured. As a result, robot applications have additional safety features, such as:

1. Perimeter guarding that when opened stops robot movement
2. Pressure-sensitive mats that inhibit robot activity if someone steps on the encircling mat
3. Vision and light beam actuators that signal the robot to stop if someone breaks the light beam or is within a sensor's vision

Robotics is an application-dependent business. That is, businesses typically don't purchase robots for the sake of having and using robots. Robots are generally purchased where applications are real, the environments are hostile, strenuous, repetitive, or dull, and the economic pressures to perform are high. Typical robot applications include spot welding, spray painting, pick-and-place material handling, gluing, and others (Figures 13-15 and 13-16). Varying degrees of robot versatility are required to suit the application. Two factors that determine the degree of robot versatility are:

1. Maneuverability of the robot arm
2. Flexibility of the control system

Many robot applications require the maneuverability of six axes of motion, as identified in Figure 13-17. The robot's position of a given point is guided to an end

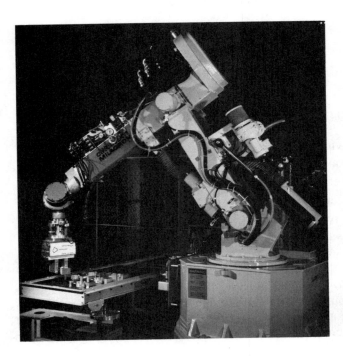

Figure 13-15 Robot pick and place application. (Courtesy of General Dynamics Corp., Fort Worth Division)

Figure 13-16 Robot performing an assembly operation. (Courtesy of Westinghouse Automation Division)

effector, for example, the center of a pair of gripper jaws or the tip of an arc welding gun. This end effector programmed point is known as the tool center point or TCP. Basically, the robot arm and its axes of movement are a means of moving the TCP from one programmed point in space to another. The control philosophy of the robot is built around the tool center point concept. The control directs the move-

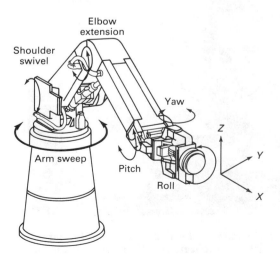

Figure 13-17 The six primary axes of robot movement. (Courtesy of Cincinnati Milacron)

ment of the TCP in terms of direction, speed, and acceleration along a defined path between consecutive points.

The primary function of robots in flexible cells and systems is to load and unload parts and tools. Although somewhat limited by part weight and configuration, robots are generally used as a material-handling device in rotational part cells. Rotational parts tend to be easier to orient in machine tools and in-process or postprocess gaging than prismatic parts. Also, rotational parts may tend to vary less in a dedicated cell, where part type, size, weight, and configuration would be relatively constant.

Robots can be interfaced with machine tools (Figure 13-18) for automated and progressive part loading and unloading. Conveyor systems can also be included to handle incoming and outgoing workpieces (Figure 13-19), along with interfacing robots with in-process and postprocess gaging systems for part checking and vision systems for optical part recognition.

Other functions that robots are being used for in cells and systems include:

1. Loading and unloading parts on transporter devices or pallets in the staging area of an FMS
2. Loading and unloading new or replacement tools in the tool-storage magazines of machine tools
3. Deburring and cleaning completed parts

Figure 13-18   Robots can be interfaced with machine tools for part loading and unloading in cellular applications. (Courtesy of Cincinnati Milacron)

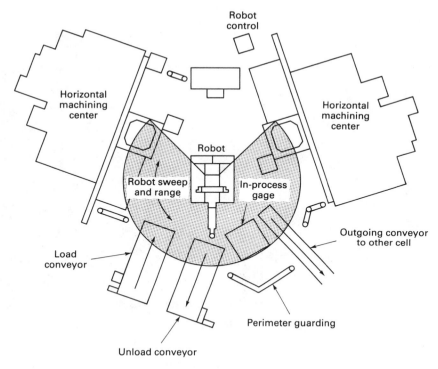

Figure 13-19   Robots can be interfaced with machine tools, conveyors, and gaging devices for part loading, unloading, and inspection in flexible cells and systems.

The industrial robot continues to be an important element to automated manufacturing. Its uses are expanding and its acceptance is gaining. In the years ahead, robots will take on broader roles in more diverse manufacturing industries and businesses.

AUTOMATED STORAGE AND RETRIEVAL SYSTEMS

In the 1950s, a revolutionary concept in material handling was pioneered in the United States called automated storage and retrieval. At the time, this concept of high-rise, high-density storage and retrieval was considered a radical change in inventory management and control, rather than a revolutionary breakthrough.

Automated storage and retrieval systems, commonly referred to as ASRSs, are automated inventory-handling systems designed to replace manual and remote-control systems. Typically, they contain tall, vertical storage racks, narrow aisles, and stacker cranes and are coupled with some type of computer control, as seen in Figures 13-20 and 13-21. For the most part, ASRSs are strictly warehouse tools that track incoming material and components, store parts, tools, and fixtures, and retrieve them when necessary.

Figure 13-20   Typical ASRS. (Courtesy of Jervis B. Webb Co.)

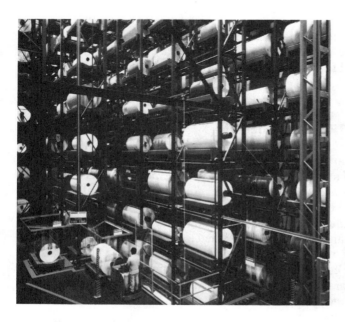

Figure 13-21   An ASRS can be used to automatically store and retrieve many different types of material. (Courtesy of Jervis B. Webb Co.)

Figure 13-22   Material is picked up at an ASRS input–output station by the stacker crane, stored in a computer assigned location, and retrieved when required. (Courtesy of Jervis B. Webb Co.)

The goal of an automated storage and retrieval system is to deliver the right material to the right place at the right time. Material is held in storage and then issued to the point of use as close to the time of use as possible.

Automated storage and retrieval systems store standard-sized pallets of material, and they have aisles that divide the storage racks. In each aisle is an arm or crane, sometimes known as a stacker crane. The crane picks up a load from an input station, stores it in a computer-assigned location, and delivers it to an output station, as seen in Figure 13-22.

Stacker cranes are rated in terms of vertical and horizontal movement in feet per second. Cranes are capable of simultaneous vertical and horizontal movement. Loads must be presented for crane movement within size, weight, and stability limits. Even if a load is within size and weight limits, an off-center load, for example, can still jam the crane.

The principal benefits of automated storage and retrieval systems are:

1. Improved inventory management and control
2. Reliable and immediate delivery
3. Space efficiency

4. Simplified and faster inventory response
5. Ability to operate in adverse environments
6. Closed storage area to reduce pilferage
7. Reduced lost or misplaced parts, tools, and fixtures
8. Design flexibility to accommodate a wide range of loads
9. Reduced labor costs
10. Accurate inventory and load location
11. Inventory reductions (as a result of improved accuracy)
12. Increased utilization potential
13. Reduced scrap and rework (resulting from manual part movement damage)

Over the past several years, many companies have focused their attention on automated storage and retrieval systems as a means to solve existing warehouse problems. In many cases, implementing an ASRS and other material movement systems have successfully reduced operating costs and gained control over the storage and retrieval process. However, business conditions and objectives determine the need for an ASRS, and a fully automated warehouse may be inappropriate for some businesses.

Overall work flow and manufacturing processes must be clearly understood and known in order to determine part, tool, and fixture movement frequency, as well as maximum and minimum load sizes. Flow to and from the areas that the ASRS is to serve should be kept clean and free of obstructions and waiting to be moved components.

The success of an ASRS, in many cases, is measured in terms of throughput. Throughput in an ASRS is a combination of many factors. These include:

1. Crane speed
2. Percent of utilization of the storage racks
3. Speed and efficiency of the AGVs or forklifts handling input and output
4. Arrangement of storage items
5. Mix of input and output operations
6. Speed of the computer controlling the ASRS
7. Open input–output points

A just-in-time strategy is often thought to be in conflict with an ASRS. One might ask, if one of the objectives of JIT is to reduce inventory, why should a high-tech ASRS be purchased to store inventory? The JIT stockless production concept recognizes that some minimum level of inventory must be carried simply to satisfy a company's marketing strategy and existing operating conditions. JIT and ASRS can and should work hand in hand to provide for peak efficiency to help reduce work in process and close the gap in a just-in-time strategy. An ASRS system actually supports JIT by getting the right parts and fixtures to the right place at the right time through controlled flow storage and retrieval and accurate inventory management.

Automated storage and retrieval systems play a major role in automated manufacturing. In an FMS, an AGV system, for example, may be linked with an ASRS to carry work back and forth between the FMS and the ASRS. In some larger installations, the FMS may have its own satellite ASRS, with a staging area shared by the two systems.

Many companies have come to the realization that to remain competitive and profitable, they must produce their products at a higher rate with reduced inventory and labor. A key advantage of ASRS is that it is a mature technology today that can fit into evolving and automating facilities of tomorrow. And, because of their greatly increased accuracy and capabilities, they can be interfaced with other equipment like robots, AGVs, and conveyors, which can add considerably to cell and FMS productivity and success.

CONVEYORS AND PALLET FLOTATION SYSTEMS

Workpiece material-handling systems are selected for automated cells and systems based on workpiece:

1. Type, size, weight, and configuration
2. Volume and throughput requirements
3. Machinery cycle time

These principal factors dictate what type of material-handling system will best suit the application and volume requirements intended. However, in many cases, the selection decision is often clouded and not that easy to make because of the variety of available comparable and equally performing material movement systems.

Conveyors offer automated manufacturing users a variety of options from which to choose, depending on individual part characteristics and production requirements. They present a hardware-defined fixed path over which components travel to their destination. Conveyors are generally classified as either overhead mounted (Figure 13-23) or floor mounted (Figure 13-24). Overhead conveyors may be either of the monorail power and free type or overhead chain type. Both power and free and chain-driven conveyors can handle medium to large part types such as automobile frames and bodies.

Floor-mounted conveyors are classified as chain, roller, or belt driven. Chain and roller conveyors are very practical and can accommodate varying types of loads (Figure 13-25 on page 273). They are sometimes used with free-floating pallets in dedicated manufacturing systems or in group technology cells where workpieces are passed on pallets from one workstation to another. Carrousel conveyors (Figure 13-26 on page 273) are sometimes belt driven and typically handle small parts. They are often used in loop applications in manufacturing cells to present workpieces to a robot for transferring to a machine tool (Figure 13-27 on page 274).

Conveyor applications are not without their inherent drawbacks and difficulties. Overhead conveyors, for example, make good use of air space but are

Figure 13-23   Overhead mounted conveyor. (Courtesy of Jervis B. Webb Co.)

Figure 13-24   Floor-mounted conveyor. (Courtesy of Jervis B. Webb Co.)

Figure 13-25 Roller conveyors are very practical and can be used in a variety of applications. (Courtesy of Eaton-Kenway, Inc.)

Figure 13-26 Carrousel conveyors are used for small parts in one- or two-machine cells. (Courtesy of Jorgensen Conveyors, Inc.)

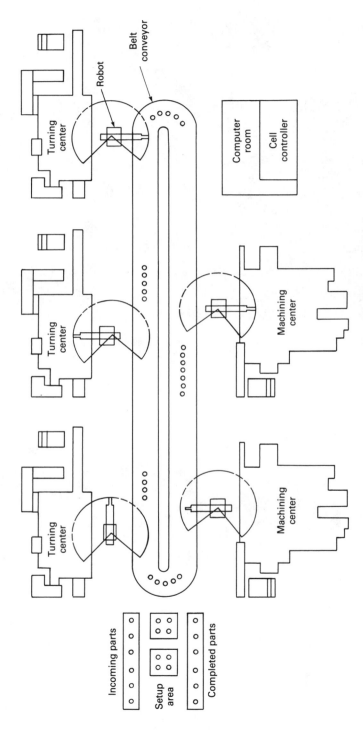

Figure 13-27 Belt conveyors are often used in cellular loop applications to present workpieces to a robot for transferring to a machine.

Figure 13-28 Pallet flotation systems are used in dedicated manufacturing systems where parts are large, bulky, and very heavy. (Courtesy of Cincinnati Milacron)

difficult to service and maintain. Floor-mounted conveyors restrict access to equipment for maintenance, support, and general cleaning. Both overhead and floor-mounted conveyors contribute substantially to shop noise levels. This can prevent operating personnel from hearing for their own safety and detecting early equipment on tooling problems.

For many applications, particularly with small parts, a choice may have to be made between an AGV or a conveyor system. AGVs have great flexibility but are inefficient users of time. Conveyors, on the other hand, have limited flexibility but are efficient for high-volume, limited-variety applications where high throughput must be maintained.

Pallet flotation systems are primarily used in dedicated manufacturing systems where individual parts are large, bulky, and very heavy. Pressurized air or liquid (usually a water-based coolant) is forced through the pallet, creating a fluid-bearing surface or film that supports the load and permits its being shuttled into the next station with minimal frictional force (Figure 13-28). Liquids are usually used for very heavy palletized parts, while air is the primary choice for lighter loads in dedicated systems.

QUEUING CARROUSELS AND AUTOMATIC WORK CHANGERS

Queuing carrousels and automatic work changers are typically part parking lots that can hold six, eight, twelve, or more pallets or workpieces in various stages of

Figure 13-29   Queuing carrousels are generally used as internal part queuing for an FMS. (Courtesy of Remington Arms)

completion. Each pallet may contain a tombstone fixture with one or more parts for processing within a cell or FMS. Queuing carrousels hold parts in queue that:

1. Are scheduled to be processed and are waiting for an open machine in order to begin processing
2. Are at some stage of processing completion and are waiting because a required machine is unavailable
3. Have completed processing and are waiting to be unloaded

Queuing carrousels are generally used in flexible manufacturing systems and are used as internal queuing for the entire system and not on an individual machine basis (Figure 13-29). AGVs typically link up to a queuing station just like the pallet shuttle mechanism on a machine tool for transferring fully fixtured pallets back and forth. Pallets are usually randomly accessed in a queuing carrousel and are not deliberately routed to a parking area. They are directed there by the FMS scheduling software based on part completion due dates and current processing situation conditions.

Automatic work changers are used as a machine-dependent part parking area for one specific machine (Figure 13-30). They typically will hold six or more fully fixtured pallets in queue. Automatic work changers are attached to the front end of a machine tool, and each individual pallet takes its turn being sequentially indexed into the machining zone of the machine tool. Automatic work changers can hold

Figure 13-30 Automatic work changers attach to the front end of a machine tool and are used as a machine-dependent part parking area for one specific machine. (Courtesy of Kearney and Trecker Corp.)

several hours of machining work and are typically purchased as part of a single machine cell. Individual pallets can then be loaded with enough work to allow the machine to operate unattended during second and third shifts.

COOLANT AND CHIP DISPOSAL AND RECOVERY SYSTEMS

Metalworking management in the past has often considered machine coolants and their disposal along with chip removal as a necessary nuisance to be controlled largely by plant laborers and maintenance personnel. However, new thinking has recently emerged brought about primarily as a result of three important factors:

1. Passage of federal, state, and local laws legislating stringent and costly procedures for disposal of hazardous waste materials
2. Introduction of modern and efficient coolant recycling and recovery systems
3. Enhanced salvage value of segregated metal-cutting chips

This change in thinking relative to waste material management and removal can yield thousands of dollars in direct annual savings because recovering and recycling used and contaminated coolant is generally more economical than disposal and replacement. As with any capital investment though, waste material removal and recycling systems have a higher initial cost, but have correspondingly greater performance, along with producing both direct and indirect savings.

Extensive studies have been conducted as to why coolants break down and

Figure 13-31   Installation and use of a high-quality coolant recovery system can reduce coolant contamination, machine clean-out time, and machine downtime. (Courtesy of Remington Arms)

fail. These failures have resulted in several causes being identified, quantified, and well documented. In general, coolants fail and require disposal or recycling because they:

1. Become contaminated with materials such as very fine metal particles, abrasive grains, and bonding material from grinding wheels and hydraulic and lubricating oils
2. Are bacterially degraded due to rapid growth of aerobic microorganisms, which turns the coolant rancid and prevents the fluid from performing its principal functions of lubrication and corrosion prevention
3. Lack the proper water–coolant concentrate mix and control

The installation of a high-quality coolant recovery system (Figure 13-31), including efficient sump cleaning, sharply reduces coolant contamination along with machine clean-out and downtime. Central coolant and chip disposal and recovery systems are used with many automated cells and systems, thereby avoiding individual machine chip and coolant service. Machines are placed over the central flume troughs (Figure 13-32) so that chips and coolant may fall directly into the trough and be swept away to a central coolant tank and chip-removal area.

Figure 13-32   Machines are placed over central flume troughs for coolant and chip recirculation and collection. (Courtesy of Remington Arms)

In central coolant and chip disposal and recovery systems, the main coolant tank, for example, will typically hold 9000 gallons of coolant and may be below ground level and outside the building housing the cell or FMS. Coolant is pumped through a central flume system at an extremely high rate, sometimes as high as 2700 gallons per minute, in order to flush chips through the flume system to the chip conveyors (Figure 13-33).

Dual flume troughs may be used in some cases for segregating ferrous and nonferrous chips. Chip diverters in the flume trough are used to programmatically direct the ferrous or nonferrous material to separate collection bins. This avoids mixing various metal-cutting chips together, thereby enhancing their overall salvage value.

Coolant recycling and recovery systems are also available for smaller-scale cells, systems, and stand-alone machines. Some machine tool transporter systems are used on an individual machine basis to replace used with recycled coolant and transport contaminated coolant to the recycling station.

The central coolant and chip-removal system is just one type of design for a cell or FMS. Some cells and systems use the traditional pick-up and delivery method of chip and coolant removal and transport for each machine. In some FMS systems, conventional chip conveyors are placed at each machine tool (Figure 13-34). AGVs are dispatched to transport loaded chip bins to a central or segregated collection and compaction area for disposal.

Figure 13-33 Coolant is pumped through the coolant troughs and central flume system at extremely high rates in order to flush chips and coolant through the flume system. (Courtesy of Remington Arms)

Figure 13-34 Conventional chip conveyors are used in stand-alone machine, cell, and FMS applications. (Courtesy of Jorgensen Conveyors, Inc.)

ANCILLARY SUPPORT EQUIPMENT

In many cell and FMS applications, special support equipment may be required to enhance or augment overall system performance and productivity. Support equipment is typically not considered to be one of the system's major elements or components but may add considerable functionality and capability to a cell or FMS.

Specialized support equipment, for the most part, can be custom tailored to suit a specific application need within a cell or system. In many cases, support equipment may be used to bring material into an operator's work zone, hold and orient a workpiece for a robot weld operation, and perform automated part holding, handling, and orientation for part or serial number marking and various other special cell or system functional support requirements.

Ancillary support equipment may or may not be classified as unique, special, or auxiliary to a particular cell or FMS depending on the type and function of the equipment and the application intended. However, some of the more commonly used cell and system support equipment are:

1. Tilt Tables

Sometimes referred to as robotic positioners, tilt tables come in a variety of types, sizes, and axis of rotation capabilities (Figures 13-35 and 13-36). Tilt tables and positioners offer one, two, or three additional axes of freedom for positioning

Figure 13-35   Tilt table or robotic positioner. (Courtesy of K. N. Aronson, Inc.)

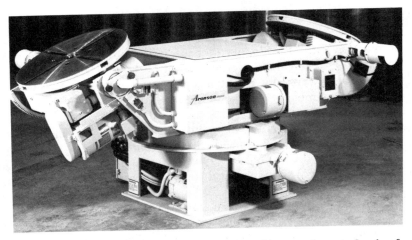

Figure 13-36 Tilt tables come in a variety of types, sizes, and axis of rotation capabilities. (Courtesy of K. N. Aronson, Inc.)

and orienting a workpiece, and many can hold and accurately position and orient loads in excess of 6000 pounds. Tilt tables are frequently used in robot welding cells to position and orient a fabrication for a variety of welds in a single clamping. Tilt tables are also used in a manufacturing cell or FMS to rotate a high tombstone fixture with one or more parts from a vertical to a horizontal position and back again for ease of operator part loading and unloading (Figure 13-37). A variety of motor and electrical options are available to assure compatibility with a system supplier's programmable controllers. As an added element of safety, pressure-sensitive mats or grating are sometimes used to inhibit tilt table movement if operating personnel are too close.

2. Robotic Transporters

Robotic transporters permit a mounted and interfaced robot to be moved and positioned back and forth on a fixed in-line track (Figure 13-38). Robotic transporters are interfaced with a robot and CNC machine tools, for example, to transport and accurately position workpieces from one machine tool to another for various machining operations. Transporters typically have total flexibility such that they can move to any workstation, in any order, in either direction along the track. Many are rubber-wheeled vehicles that mechanically lock to the track at workstations, enabling the tallest robots to safely extend their arm. Many transporters are capable of a 300 foot per minute traverse rate; they are equipped with numerous safety devices because of their high-speed capabilities.

Figure 13-37   Tilt tables are used in cells and systems to make part loading and unloading easier for the operator. (Courtesy of Cincinnati Milacron)

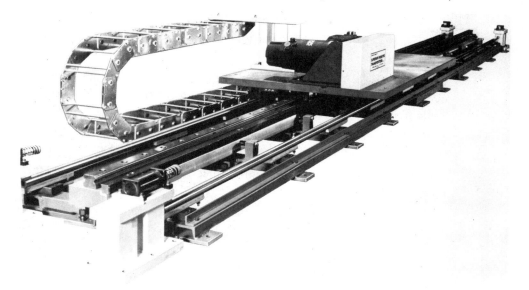

Figure 13-38   Robotic positioners permit a mounted and interfaced robot to be moved and positioned back and forth on a fixed in-line track. (Courtesy of ICL Manufacturing)

SUMMARY

1. AGVs are bidirectional, battery-powered, driverless vehicles that can be programmed for path selection and positioning and are equipped to follow a changeable or expandable guidepath.

2. AGV guidepaths must be laid out to access all processing and queuing stations as part of the FMS master plan.

3. Robots are reprogrammable multifunctional manipulators that move material, parts, and tools.

4. Robots are classified as either nonservo or servo and are application dependent.

5. The primary function of robots in flexible cells and systems is to load and unload parts.

6. Robots are classified by their arm configuration and reach, power sources and speed, load capacity, application capabilities, and control techniques and intelligence.

7. Automated storage and retrieval systems (ASRSs) contain tall vertical storage racks, narrow aisles, and stacker cranes and are coupled with computers for automatically storing, tracking, and retrieving material.

8. Implementing an ASRS can successfully reduce operating costs and gain control over the material storage and retrieval process.

9. The success of an ASRS is measured in terms of throughput.

10. An ASRS supports JIT by getting the right material to the right place at the right time.

11. Conveyor systems are classified as either overhead or floor mounted; overhead conveyors may be either of the power and free or overhead chain type, while floor-mounted conveyors may be chain, roller, or belt driven.

12. Queuing carrousels and automatic work changers are part parking lots that hold parts waiting to be machined, unloaded, or reprocessed.

13. Coolant recycling and chip collection and salvage can yield thousands of dollars in direct annual savings.

14. Support equipment, in most cases, is custom tailored to suit a specific application need within a cell or FMS.

REVIEW QUESTIONS

1. The importance of support equipment to flexible cells and systems increases in value as the size and complexity of the system increases. True or False?
2. Name the seven types of AGV guidance systems.

3. Which of the following is not an advantage of AGVs over conventional material-handling systems?
   (a) Better resource utilization
   (b) Reduced product damage and less material movement noise
   (c) Ability to move material faster
   (d) High location and positioning accuracy
   (e) Increased throughput

4. An AGV traffic control system must include the capability for automatically tracking the current status of all payloads active in the system. True or False?

5. An AGV _____ must be laid out as part of the FMS master plan.

6. Industry standards for classifying all robots have been established by the Robot Institute of America. True or False?

7. Fixed robot mounting types consist of floor mounted, machine mounted, wall mounted, and _____.

8. Name the six axes of robot motion.

9. The primary function of robots in flexible cells and systems is to _____.

10. The primary goal of an ASRS is to:
    (a) Control inventory
    (b) Store material, tools, parts, and fixtures
    (c) Deliver the right material to the right place at the right time
    (d) Make more efficient use of storage space

11. Installing an ASRS should reduce scrap and rework resulting from part movement damage. True or False?

12. A method of measuring throughput success in an ASRS is by the mix of input and output operations. True or False?

13. Explain how an ASRS supports a just-in-time manufacturing strategy.

14. Name the types of floor-mounted conveyors.

15. Both overhead and floor-mounted conveyors contribute to shop noise levels. True or False?

16. _____ are used as a part-holding buffer storage area for an entire FMS.

17. Installing a central high-quality coolant recovery and recycling system will:
    (a) Reduce coolant contamination
    (b) Produce direct annual savings
    (c) Reduce machine clean-out and downtime
    (d) Eliminate water–coolant concentrate mix problems
    (e) All of the above
    (f) Only (a), (b), and (c)

14

Cutting Tools and Tool Management

OBJECTIVES

After studying this unit, you will be able to:

□ Understand the importance of cutting tool control, preset, identification, and monitoring to a successful FMS installation.

□ List and describe the four required elements of a tool management system.

□ Name the four most commonly used tool strategies as applied to tool management.

□ Identify some important factors to consider relative to cutting tool deformation and wear.

INTRODUCTION

Some of the most cumbersome and difficult issues to deal with relative to FMS are managing, coordinating, and controlling the wide variety of cutting tools. This includes not only having and maintaining the required number of cutting tools to process the required parts through the FMS, but also managing and coordinating other elements, such as redundant or replacement tools, tool assembly component requirements, tool storage, reconditioning and preset considerations, tool life monitoring, broken tool detection, and a host of other factors.

286

In many cases, too much variety exists; the number of different tools required to machine the identified parts has proliferated because of a lack of basic tool control prior to FMS. Through careful and astute manufacturing analysis of each workpiece, minimal design engineering changes and/or optimized part programming routines may reduce or eliminate unnecessary tools. Analysis of this type for each part can have a considerable effect on reducing overall cutting tool requirements for parts run through the FMS.

Regardless of how well an FMS may operate or perform, "it's what's up front that counts." And in the area of machining, the cutting tool is "up front." Managing what's up front, the cutting tools, means getting the right tool to the right place at the right time. Without the existence of high-quality, well-maintained cutting tools in the framework of a comprehensive tool management system, the productivity of an FMS can be severely limited.

Typically, tooling schemes in industry today do not operate with the discipline required by FMS. And, as a result, past tooling inefficiencies tend to remain unchanged, considerably impairing the efficiency and productivity of the FMS. Departure from these traditional schemes presents opportunities for significant savings.

GETTING CONTROL OF CUTTING TOOLS

Before simply implementing an FMS and using or adopting existing cutting tool practices and policies, examination of the current practices, procedures, and methodologies is in order. Controlling the cutting tools involves good tooling policies, cost-effective part programming strategies on the machine, and sound tool-related practices in tool rooms, manufacturing engineering, and other off-line operations.

By neglecting to review existing cutting tool practices and procedures with an incoming FMS installation, management loses a valuable cost-saving and optimization opportunity. FMS helps to force such a reexamination of existing policies and procedures because many routine practices will have to change anyway to accommodate the FMS, and perpetuating inefficient cutting tool practices contributes to less than optimum overall manufacturing efficiency.

In general, gaining control of cutting tools, which can realistically be applied independent of FMS, should begin as part of the overall FMS preparation plan long before any equipment is installed. The following items should be considered as cost-effective, optimizing tactics to begin to augment and enhance the full impact of FMS productivity effectiveness.

1. Review cutting tool and indexable insert inventory and get control of usage. Existing perishable cutting tool and insert inventories should be reviewed and categorized in terms of use. Excess quantities should be purged from inventory (sold) and their value recovered if possible. Records should be kept with respect to frequency of use in order to make intelligent quantity and cost buy decisions.

2. Review cutting tool purchasing practices. In many cases, cost-favorable buy decisions are made because of quantity discounts. However, overall use may take several years to deplete. Consequently, purchasing policies and practices may require tighter controls and more rigid guidelines tied closer to actual verified use.

3. Reduce dependency on specialized, nonstandard tooling. This practice begins with disciplines imposed in design engineering. Joint efforts between design engineering and manufacturing engineering should be undertaken to promote and use standard cutting tools whenever possible. Emphasis should be placed on "design for manufacturability" centered around producability within a defined set of standard tools and discouraging the need for special, one-of-a-kind tools and tool purchases.

4. In some cases, as perishable tools are checked out from in-house tool stores, they ultimately find their way into operators' tool cabinets. Carried to extremes, such out of control activity can resort to unnecessary and costly extra perishable tool purchases. Such operator "hording" can extend to tool holders, adapters, and collets, as well as other tool assembly components. Periodic management tool searches may be required to return tools and tool components to their central storage facility. On the other hand, operator hording could be an indicator of tool store quality, turnaround, or availability problems.

5. Guidelines for tool assembly preparation and reconditioning (Figure 14-1) must be established and enforced. In some cases, cutting tool assembly and preset personnel may make discretionary substitutions of specific type, diameter, and length of tools. "Free substitution" of tools can be an invitation to disaster at the machine tool, along with the potential for personal injury. Substituting tools can further cause incorrect tool ordering, replacement, and stocking. If correct tools are unavailable, programming should be contacted for substitutions. Tooling changes, modifications, or substitutions should only be initiated by manufacturing engineering and programming personnel. Reconditioning should also follow strict guidelines with respect to how much of the drill, end mill, and so on can be reconditioned. Cutting tools must maintain specified flute lengths, point angles, and a number of other factors with respect to tool assembly drawings (Figure 14-2) that the part programmer is working to. As a result, tool reconditioning and sharpening or assembly must not take actual tool dimensions out of tolerance with respect to stated dimensions that programming personnel rely on. Tool assembly dimensional standards must be rigidly enforced.

6. Programming personnel must also work to impose guidelines and restrictions relative to tool proliferation. In some cases, for example, NC programmers add additional tool assemblies to the existing tool library because of minor differences in tool lengths required to meet specified print dimensions. Adding unnecessary extra tool assemblies to the overall tool assembly library will cause extra tool holders, collets, adapters, and the like, to be purchased. These support the additional perishable tool purchases, thereby contributing even more to tool inventory costs. And adding additional tool assemblies means extra tool pockets used and more tool change time. In some cases, requesting design engineering to change the depth of a hole, for example, will permit the use of an already existing tool

Figure 14-1   Policies, procedures, and guidelines for cutting tool assembly and reconditioning must be established and rigidly enforced. (Courtesy of Remington Arms)

1.090 S.S. boring bar

Tool no.
55109436

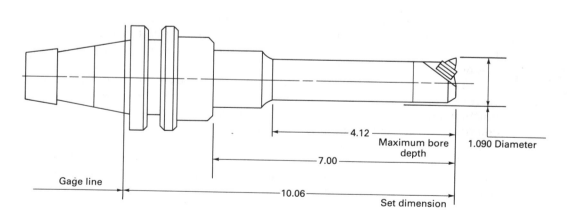

4.12

Maximum bore depth

1.090 Diameter

7.00

Gage line

10.06

Set dimension

Tool assembly no.
55109436

Figure 14-2   Cutting tools must maintain specified lengths (set dimensions), flute lengths, diameters, and other dimensional data with respect to tool assembly drawings.

Figure 14-3   Some machining centers can contain a numerous variety and supply of cutting tools. (Courtesy of Cincinnati Milacron)

assembly to be used without adding an extra tool. Tooling variety and proliferation can undermine an FMS. Typically, this is the result of traditional approaches to process planning and part programming. Every effort should be taken to avoid adding additional tools to the library of tool assemblies.

Today, some of the best planned FMSs have a week of cutting tool capacity in each tool magazine (Figure 14-3). Such lack of attention to existing cutting tool practices and policies prior to FMS can also lead to questionable FMS reliability, compromised flexibility, and added problems for integration of system elements. A well-coordinated program of cutting tool management, use, distribution, and cost control, begun well in advance of FMS delivery, can have a significant impact on reducing cutting tool use, consumption, and inventory costs while improving overall operating efficiency and utilization levels.

TOOL MANAGEMENT

Regardless of how "flexible" a flexible manufacturing system (FMS) is, the system is still only capable of processing a finite number of parts. The overall flexibility or "randomness" of an FMS is typically constrained by two support resources:

palletized fixtures that control the rate of work flow coming into and going out of the system, and tool storage capacity. However, with an acceptable tool management philosophy and tool management system, tool capacity constraints can be considerably reduced.

The main problems caused by tool capacity constraints and a lack of tool management are:

1. Insufficient redundant tool backup at the machine in case of tool breakage or tool wear conditions
2. Automated tool change taking place at the spindle, but manual tool change at the matrix when tools require maintenance or exchange
3. Insufficient use of preset tools and excess tool inventory
4. Conflicting priorities with other areas outside the FMS over tool availability and reconditioning
5. A limited number of workpieces being available to process due to insufficient tool matrix capacity
6. Underutilized machines or low production rates caused by too many tools and extensive tool changing

Generally, tool management is getting the right tool to the right place at the right time. This is extremely important in automated manufacturing and allows the desired part mix and quantities to be manufactured while maintaining acceptable system utilization and minimizing work-in-process inventory.

Having an acceptable tool management system to fulfill the tooling requirements of an FMS means adequately addressing the following four areas:

1. Tool Room Service

Providing adequate tool room service does not directly affect the machine tools that make up an FMS, but is essential to system effectiveness. Tool room service (Figure 14-4) is a necessary support function, dealing principally with preparing, servicing, organizing, and controlling the vast array of perishable tools, inserts, tool holders, and tool components. Information to build tools is based on the part program requirements of the designated part spectrum for a given production run. Tools must be accurately prepared, inspected, and gaged in a timely manner in order to meet the stringent FMS production scheduling requirements. Strong inventory and management control practices must be applied in order to make the FMS tool room perform with predictability and precision. This ranges from monitoring, controlling, and purchasing the large number of perishable and durable tool components to hiring, training, and effectively managing the tool room function. Additionally, some type of automated delivery of cutting tool assemblies, including redundant and replacement tools, must be adopted to support the FMS part flow and production rates. The principal elements of tool room service are:

- ☐ Buildup and teardown of tool assemblies
- ☐ Inventory of tools, tool components, and related tool assembly instructions
- ☐ Control of idle (returning and least used) tool assemblies along with determining tool disposition (what to do with these tools)

Figure 14-4   Tool room service is a necessary support function for tool management. (Courtesy of Remington Arms)

□ Actively maintaining NC tool data for the remaining tool life of returning or idled tool assemblies; this is important, for example, in the case of a returning tool assembly where adequate tool life remains; knowing how much tool life remains can divert the tool assembly to another tool group for manufacturing another part and prevent the teardown and buildup of the same or a duplicate tool

2. Tool Delivery

Tool delivery addresses the tool management function relative to moving the tools between the tool room and the various tool magazines of each machine tool in the FMS. This includes transporting the tools to and from the machine tool requiring those tools, and loading and unloading the tool magazine once the tools arrive at the machines. If the demand for tools based on the variety of part mix is high enough, complete automation of the tool delivery and distribution function may be necessary, as seen in Figure 14-5. An automated tool receiver unit receives its commands to operate (which tools to remove and which ones to add) from a programmable controller, which receives its information from the FMS host computer. If this is required in the FMS, then additional hardware and software interfaces will also be required. Automated tool delivery optimizes FMS effectiveness by allowing the tool delivery mechanism (Figure 14-6 on page 294) to exchange tools at the rear of the machine tool while the machine stays in production. This is extremely important to maintaining the spindle utilization levels required to run the FMS at optimum production rates.

Figure 14-5   If tool demand is high enough, complete automation of the tool delivery function may be necessary. (Courtesy of Kearney and Trecker Corp.)

3.   Tool Allocation and Data Flow

Tool allocation and flow are two of the most difficult aspects of tool management to manage and control. Tool allocation is essentially assigning and controlling the total number of tools required for each machine to process the previously defined FMS part spectrum. It is based on specific part process plans, NC programs, and machining methodology, along with the varying part mix and volumes that could be running through the system at any given time. Determining such tool allocation requirements is easy as long as infinite tool inventory, infinite tool capacity at the machine, and error-free, moments-notice tool room service is assumed. However, even though these are assumptions, attempting to operate an FMS without attaining some level of achievement of these assumptions constrains FMS spindle utilization and overall system effectiveness. Various approaches to overcoming this complex problem will be discussed later in this chapter. Controlling the tool data flow relative to the allocated tools requires that the MCU (machine control unit) receive data about each tool at the machine. This would assume tool data transfer from the preset area as tools are automatically gaged, identified, and entered into the FMS tool system data base. Tool data transfer is discussed in more detail under the Tool Preset Identification and Data Transfer heading.

Figure 14-6   Automated tool delivery optimizes FMS effectiveness by allowing the tool delivery mechanism to exchange tools at the rear of the machine without interrupting chip making. (Courtesy of Kearney and Trecker Corp.)

4. Fault Sensing

Fault sensing is monitoring and detecting cutting tool problems at each machine. This involves electromechanical or optical sensing or detection of worn or broken tools along with tool absenses or misplacements. Each tool is offset to a contact or noncontact sensor (Figure 14-7) each time it is used in order to validate tool presense, correctness, and condition. If the tool is broken or the tool life has expired (discussed later in this chapter), replacements should be available in the tool magazine. If any tool fails for any reason, a replacement must be obtained from the tool room in a timely manner. The ability to sense and respond to tool fault conditions is an important issue in FMS tool management.

Tool management, as previously mentioned, is one of the most difficult aspects of FMS to regulate and control but also one of the most vital. The information just discussed describes what tool management is. The following discussion describes the most used and considered tooling strategies for FMS applications.

TOOL STRATEGIES

Various tool strategies exist within the framework of tool management that require examination. Each has its advantages and disadvantages, as well as particular

Figure 14-7 Cutting tool fault sensing can be detected by offsetting each tool to a contact or noncontact sensor in order to validate tool presence, correctness, and condition. (Courtesy of General Dynamics Corp., Fort Worth Division)

application for an FMS. Although other tool strategies may be available, the following are the most accepted and used:

□ Mass exchange
□ Tool sharing
□ Tool migration
□ Assigned tools

1. Mass Exchange

The mass-exchange strategy (Figure 14-8) is removing all the tools in each machine tool matrix at the completion of specific production requirements and replacing them with the new part required tooling. Essentially, every tool needed to manufacture each part coming to that specific machine in the FMS for manufacture must be provided. This strategy is similar to the complete tool set exchange in a job shop environment, where all the tools are emptied and replaced once batch or part lot requirements are completed. The mass-exchange strategy does not take into account the fact that tool sharing can reduce the overall tooling inventory and tool-handling time. Mass exchange permits tool-exchange control to be minimized at the expense of an increase in tool inventory. The mass-exchange strategy is logical and attractive for FMS applications only where high-volume and low-part variety workpiece requirements exist.

2. Tool Sharing

The tool-sharing concept permits the logical sharing of tools within the framework of a fixed production period and workpiece requirements. Common tooling among the fixed production requirements is recognized, identified, and shared among the various parts to be manufactured in the fixed production period. Tool

Figure 14-8   The mass-exchange strategy is removing all tools from each machine tool matrix at the completion of specific production requirements and replacing them with the next set of tools. (Courtesy of Mazak Corp.)

sharing represents a vast improvement over the mass-exchange strategy because identified tools to be shared are not duplicated in each machine's tool matrix (Figure 14-9) for each workpiece type, thereby reducing tool inventory. After fulfilling part requirements within the fixed production period, a new set of tools for the next production period is loaded and common tooling is again identified. Adopting the shared tool strategy is essentially a mass exchange within a fixed production period, rather than a mass exchange based on a specific workpiece type. Also, the shared tool strategy requires computer software to implement because of merging the tool lists and matching requirements to identify the common tooling.

3. Tool Migration

The tool-migration strategy is basically an extension of the mass-exchange and tool-sharing theory. Both the mass-exchange and tool-sharing strategies had to consider the workpiece to be manufactured within the fixed production period and the tool matrix capacity available to support it. The tool-migration strategy ignores how long a given production period is because it is not affected by the available tool matrix capacity. As various parts are completed, many tools used to manufacture those parts become available for removal from the tool matrix. Removing these tools frees tool pockets in the tool matrix and permits other tools needed for new arriving parts to be loaded. Tool-migration exchanges must be done in an effort to

Figure 14-9   Tool sharing represents a vast improvement over the mass-exchange strategy because identified tools to be shared are not dupliated, thereby reducing tool inventory. (Courtesy of Cincinnati Milacron)

minimize spindle interruption, which is of primary importance. Consequently, tools completing their manufacturing service are removed from the matrix at the rear of the machine tool and tool matrix, while needed new tools are inserted in available tool pockets. Tool delivery can be accomplished through various means. One principal way is through an AGV (automated guided vehicle) at the rear of the machine (Figure 14-10) whose arrival is timed with completing and upcoming tool-requirements. The tool-migration strategy permits reducing the tool inventory even more through sharing between various production periods. However, this strategy requires the application of sophisticated computer software and decision logic in order to determine, for example, which tools to remove from the matrix and which are needed for new part requirements. Some removed tools may also be needed at a later date, so deciding which tools to remove could affect part sequencing and the total number of tools handled.

4. Assigned Tools

The three strategies previously discussed assumed that a given set of workpieces will probably be machined at a specific machine tool. However, this is not necessarily an accurate assumption. The reality of manufacturing operations forces consideration of production schedule changes, machine breakdowns, tooling and material unavailability, and the like. Flexibility among processing equipment therefore becomes a high priority. And, in many cases, manufacturing flexibility may be dictated by system design. Machine tools may be grouped, thereby enabling any machine in the group to manufacture the part. But desiring flexibility among grouped machines requires duplicate tooling at each of the machines within the group. The assigned tool strategy can address the need for increased flexibility among a set or group of machine tools. This strategy identifies the most-used tools

Figure 14-10   Tool delivery can be accomplished through various means; one of the principal ways is through an AGV at the rear of the machine tool. (Courtesy of Kearney and Trecker Corp.)

for the production requirements and part mix and assigns permanent residence (Figure 14-11) to those tools in each machine tool matrix for the full production run. Tool migration can then take place with the remaining available pockets. However, with this strategy, minimizing tool inventory is not achieved, but the flexibility to respond to unplanned events and delays is considerably enhanced.

To better depict the effectiveness of the various strategies, two graphs (Figures 14-12 and 14-13 on page 300) are shown. Figure 14-12 shows the relationship between tool inventory and the various tool management strategies. Figure 14-13 depicts the level of control required (computer software decision logic) relative to the various strategies. In any case, the various tool strategies must be analyzed very closely in order to make the proper selection to suit a given set of FMS requirements and conditions.

TOOL PRESET, IDENTIFICATION, AND DATA TRANSFER

Although the entire aspect of tool management is not as obvious or glamorous as other working components of an FMS, it is absolutely critical to keep tools available and ''in the cut'' to obtain expected system performance levels. Tool preset,

Figure 14-11 The assigned tool strategy identifies the most-used tools for the production requirements and part mix and assigns permanent residence to those tools. (Courtesy of Mazak Corp.)

identification, and data transfer are dynamically evolving elements briefly discussed under Tool Management, but demanding closer attention.

Tool presetting can easily be performed with simple height gages or micrometers. However, this manual approach is highly labor intensive, time consuming, and, most importantly, open to human error, as touch sensitivity is highly sub-

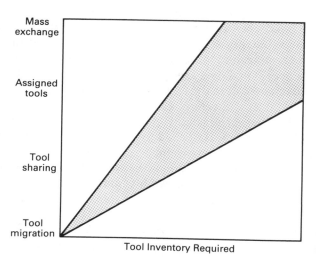

Figure 14-12 Graph depicting the relationship between tool inventory and the various tool management strategies.

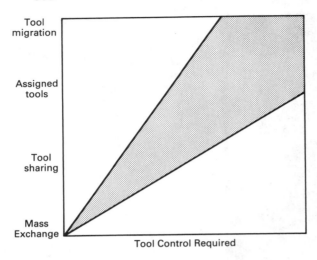

Figure 14-13 Graph depicting the level of control required (computer software decision logic) relative to the various tool management strategies.

jective. Therefore, many presetting machines are based on touch–readout tool gages (Figure 14-14) or optical projection systems that magnify the tool point (Figure 14-15). Using an optical projection system, the entire tool assembly is inserted into the presetting spindle, and the tool preset operator adjusts the slides until the tool point is aligned with the crosshairs on the projector screen. The measured X and Z coordinates are displayed on the digital readout display (Figure 14-16). Readout display information is then recorded either manually or electronically for inputting to the MCU when the tools are loaded and ready for manufacturing duty.

Optical methods of tool presetting have now been extended to machine vision (Figure 14-17). Optical machine vision systems, both accurate and repeatable within 0.0002 inches , are also easy to use and gaining wide acceptance. With such systems, tool preset information is programmed in for each tool. When a tool is measured, its code number is input, the system retrieves the tool parameters, the operator gets the tool point on the monitor, and the system is actuated to calculate the preset data. Tool and data identification is concerned with recognizing a tool's identity and making sure its dimensional data are associated with the right tool. Early machining centers called up tools from the tool drum or storage matrix either sequentially or randomly by tool pocket location. Use then centered around tool identification through an eight-digit tool assembly number. Each number pertaining to a specific sketch (Figure 14-18) was applied to a uniquely different tool. The eight-digit tool assembly number was then assigned to a specific pocket in the tool matrix by the machine operator through loading the tool and inputting the eight-digit number via the MCU keyboard. Tool assembly numbers could also be entered in the MCU automatically at the tool gage when the tool length was gaged at preset.

Sophisticated identification systems are available and are being used in FMS and other factory automation applications but have much broader and long-range

Figure 14-14   A typical contact and readout display tool preset gage. (Courtesy of Cincinnati Milacron)

Figure 14-15   Optical tool preset gage. (Courtesy of Remington Arms)

Figure 14-16   Display of cutting tool measured values. (Courtesy of Remington Arms)

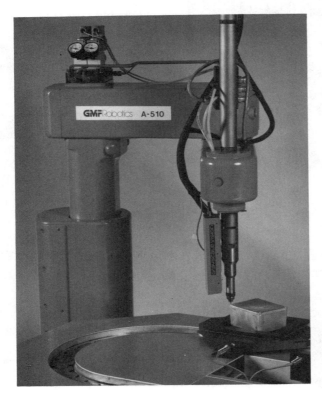

Figure 14-17   Optical vision systems are accurate, have good repeatability, and are gaining wide acceptance. (Courtesy of GMFanuc Robotics Corp.)

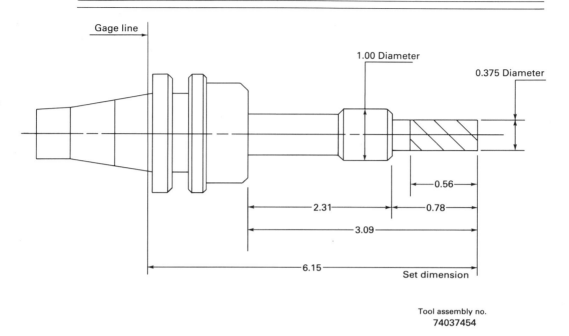

Figure 14-18 **On some stand-alone CNC machines, the eight-digit tool assembly number for each unique cutting tool is assigned to a specific pocket in the tool matrix by the machine operator.**

potential. Automated identification systems are important because they are reliable, save time, and reduce human error. The most common of these identification systems are:

1. Bar Code Scanning

Bar coding is the most popular form of automatic identification as evidenced by supermarket check-out lanes and other retail business use. It is being applied more and more throughout the manufacturing process. With NC (FMS), bar codes are imprinted on paper or Mylar and fastened to the tool holder with adhesive or engraved in the tool. The control unit remembers the pocket where each unique coded tool was placed. Bar codes (Figure 14-19) are made up of binary digits arranged so that the bars and spaces in different configurations represent numbers, letters, or other symbols, depending on the symbology used. Scanners that read bar codes (Figure 14-20) contain a source of intense light produced by a laser or light-emitting diode and aimed at the pattern of black bars and spaces of varying widths. The black bars absorb the light, and the spaces reflect it back into the scanner. The scanner then transforms the patterns of light and dark into electrical impulses that are measured by a decoder and translated into binary digits for transmission to the computer.

Figure 14-19   Bar codes are made up of binary digits arranged so that the bar and spaces in different configurations represent numbers, letters, or other symbols.

2. Machine Vision

Machine vision (Figure 14-21) is an imaging process involved with scanning and interpreting objects, documents, or labels. Although the imaging process itself is more complex than that of bar code scanning, the technology has potential for a large number of applications, many of which are FMS related. Applications would include character reading, sorting by shape or markings, locating defective parts, inspecting products, and positioning carts, parts, or pallets. Continued advances in machine vision technology and capabilities will make this method of automated identification more applicable for a variety of potential uses.

3. Radio Frequency Identification

This form of automatic identification employs bidirectional radio signals as the encoding medium and is widely used to provide hands-free access control. Radio-frequency identification offers solutions to application problems in industrial automation and material handling where there is no line of direct sight between the scanner and the identification plate or tag.

Figure 14-20   Scanners that read bar codes contain a source of intense light produced by a laser or light-emitting diode aimed at the pattern of black bars and spaces. (Courtesy of Kearney and Trecker Corp.)

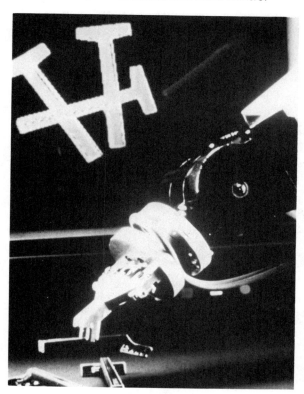

Figure 14-21 Machine vision is an imaging process involved with scanning and intrepreting objects, documents, and labels. (Courtesy of Westinghouse Automation Division)

4. Optical Character Recognition

Optical character recognition utilizes human readable letters and numerals, rather than the lines and bars of bar coding, that are scanned with a light source. When a particular pattern is recognized by the scanner, the data are converted to electronic impulses for transmission to the computer. Optical character recognition and bar code scanning have been combined to implement automation in the U.S. Postal Service.

5. The Microchip

Microchip identification employs the use of a microchip embedded in a sealed capsule that can be inserted in the tool holder. This system, as seen in Figure 14-22, uses a noncontact read-only head that can be attached to tool changers, presetting fixtures, or tool grippers. Reading can occur at a distance of up to 0.080 inch, and the read time is less than 50 milliseconds, with an allowable 0.120 inch misalignment. The microchip can also be programmed off-line with the tool identification and other dimensional data. The embedded microchip can be read by a proximity sensor to identify each tool. Although read–write microchip identification systems exist, read-only is most appropriate in an FMS or host computer application.

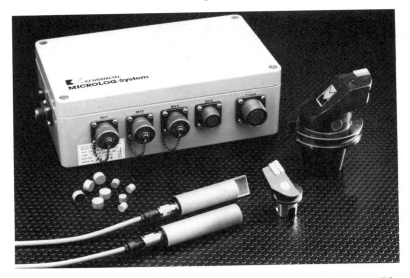

Figure 14-22 Microchip identification employs the use of a microchip embedded in a sealed capsule that can be inserted in the tool holder and read by a noncontact read-only head. (Courtesy of Kennametal, Inc.)

Once cutting tools have been assembled, gaged, and identified, the associated tool dimensional data must now be transferred to the host computer for application use. The tool preset operator assigns an identification number to the entire physical collection of tools. This identification associates the physical collection of cutting tools with the data that are collected on each of the tools. If an electronic tool gage is being used, the gaged values of tool length and diameter are automatically read from the gage and transferred and stored in a tool collection file on the FMS computer. Gaged values can also be entered manually from a tool gage terminal and uploaded to the tool collection file on the FMS computer. Upon completion of the collection of tool data values, the entire file of tool data is assigned the same identification code as that of the physical collection of tools for a specific part program. Cutting tools are now prepared and tool dimensional data transferred to the tool system data base in the FMS computer for association when those tools are called to use.

Flexibility is required in tool data transfer because the quantity of tools required to process a specific part, for example, may exceed the quantity that can be transported at one time. And multiple collections may need to be prepared in order to tool more than one machine tool to completely process the part. In such cases, separate tool collections and tool files would need to be prepared and identified based on the available conditions. The tool preset operator can use an on-line terminal (Figure 14-23) connected to the FMS host computer to display the status of various tool files and to interrogate the tool system data base to determine missing tools from incomplete collections.

Figure 14-23   **Tool preset operators can use on-line terminals connected to the FMS host computer to display tool file status and interrogate the tool system data base. (Courtesy of Remington Arms)**

TOOL MONITORING AND FAULT DETECTION

Cutting tools can be monitored in stand-alone NC, cell, and FMS applications for tool life. Tool life refers to the time during which a cutting tool produces acceptable parts in a machining operation (Figure 14-24). A cutting tool is considered to have reached the extent of its useful life when any of the following occur:

1. Loss of dimensional part accuracy due to cutter wear or deflection
2. Loss of required surface finish due to tool wear, material buildup, or chatter
3. Tool breakage or chipping due to excessive wear
4. Excessive torque from increased feeds and speeds and heavy stock removal applications

Tool monitoring therefore becomes a comparison of how much useful life should exist on a given tool measured against the actual cutting time of the tool. When the actual cutting tool time as tracked by the host computer in FMS applications expires, the FMS can be set up to perform one of the following actions:

1. Select a duplicate or redundant tool, continue operation, and notify the FMS computer
2. If no duplicate tool has been provided, stop the machine in a "feed hold" condition and notify the FMS computer
3. Flush the part requiring that specific tool from the active machine queue and go on to another part (if different)

Figure 14-24 Tool life refers to the time during which a cutting tool produces acceptable parts in a machining operation. (Courtesy of Kearney and Trecker Corp.)

Tool monitoring measures normal tool wear against a predetermined standard stored on the FMS host computer and excludes recognition and detection of major tool failures or breakage. However, because of sophisticated FMS software, look-ahead capabilities exist that can tell when time will expire on a given tool and determine how many duplicates are required. Unpredicted catastrophic tool failures can be detected by means of measuring a tool's torque at the spindle or through tool fault sensing and detection. These will be discussed in more detail later in this chapter.

An important first step of tool monitoring is to build an accurate and reliable tool data base. Because of the vast differences in users' interpretations of tool life, the gathering and compiling of machinability information for a tool life data base should be the responsibility of each user. Cutting tool vendors can supply some preliminary information based on material type, cut speed, feeds and speeds, and the like, but the best data for tool life expectancies will come from user experimentation and learn as you go estimates under actual shop conditions.

In a random-order FMS, it is possible to run parts consisting of different material types. Care should be exercised when setting up the tool life data base to categorize tools by part material type. This discipline is forced in most cases because a specific cutting tool design is best suited for cutting a specific type of material. Where one tool can be used effectively on different materials, the user should select the worst-case tool life or dedicate a specific tool for a specific material, even though this may add to the tool data base and tool inventory.

Tool life monitoring places heavy emphasis on copies of redundant tools. Enough cutting tools, related components, and holders must be available to provide for a constantly changing mix of parts to be machined. It must be remembered that some tools (assemblies) will be available for backup, some in each tool matrix at the machines, and some waiting to be disassembled or resharpened.

Two strategies relative to redundant tools to replace expired life tools are:

1. Keep all required tools, including redundant copies, present on the machine. All tools (originals plus copies) required to process those workpieces through a given period of time are present on the machine. The number of redundant copies of any tool is dictated by the total tool life required to cover the projected total cutting time (worst case) for that tool during the machining period.

2. Have only one copy of required tools present on the machine. To operate in this mode, an FMS requires that only the unique tools necessary to process a set of workpieces authorized for a machine need be present on the machine. "Extra copies" may be exchanged with expired life tools via an automated tool delivery system as required.

When operating with only unique tools in the tool matrix, having an individual tool's life expire is part of normal operating conditions. When the life of a tool expires under such conditions, the MCU must stop the machine cycle without losing part program alignment, display an alert message and notify the FMS computer, and resume when the tool has been replaced. In a completely unattended operation with all redundant tools present and available in the tool matrix, when the last copy of a tool completes its life cycle, processing stops. Suspending part processing because a single tool's life has expired would then cause major production problems as workpieces behind the affected one would be blocked. In cases such as this, the MCU must be capable of conditionally branching to a specified statement label in the NC part program, marking the workpiece as incomplete for that operation, exiting the program, and moving on to the next part in queue. If, after completing workpiece requirements, a tool returns with remaining tool life (this can be checked by the tool preset operator by comparing the time used against the time available in the FMS tool data base), it can be placed in store for job reassignment. If the tool is not required for job reassignment or does not pass the requirements for condition of tool life, it will be torn down, reconditioned if required, and built up for a new job order. In all cases, tools should be returned to the tool room for checking before reassignment.

Fault sensing and detection capabilities are designed to detect and provide recovery for major tool damage, failure, or breakage within a tool's active life cycle. Fault detection is generally provided by:

1. Adaptive or torque measurement and control
2. Broken tool sensing

An adaptive or torque measurement capability works through built-in sensors that detect when certain tools begin to draw more than acceptable levels of horsepower (Figure 14-25). For small-diameter tools for which horsepower values are too small for practical measurement, optical or length-sensing devices may be used for verifying tool length. Out-of-tolerance tool lengths are declared "broken" by the system. Torque measurement and control are essential for FMS because of its ability to adapt feed rates under varying stock removal conditions, as well as sense

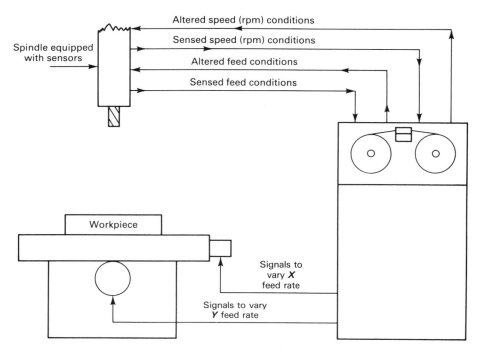

Figure 14-25   **The adaptive control or torque measurement capability works through built-in sensors that detect when certain tools begin to draw more than acceptable levels of horsepower.**

broken or dull tools during unattended operation. This feature, activated through the NC part program, can also provide the ability to sense torsional overload of the machine spindle for protection and turn coolant on and off based on tool torque so that coolant is on only during machining.

Broken tool sensing has long been considered an important feature of FMS. Detection alone without a versatile and automatic recovery strategy offers little value in an unattended system. Also, the ability to automatically or programatically gage, check, and set tool lengths is a requirement in an FMS. Both of these features are made possible using a table-mounted fixed or optical probe as seen in Figure 14-26. The fixed probe is a plunger-operated proximity switch that can be permanently mounted on the machine in a position where interference is minimal and connected directly to the MCU by a cable. The optical probe performs tool-length checking by optical means in the same manner based on where the tool tip should be as compared to the known tool length. Tool-length checking (verification that the tool is not broken off) and tool-length setting (resetting a tool length) may be used at any time in the part program to offset the cutting tool to the fixed or optical probe for checking or setting.

Upon detection of a broken tool, the MCU completes the cycle, putting the tool back in the matrix, and marks the tool as "broken" in the MCU's current tool file, searches for a redundant copy, and, if found, offsets to the fixed probe for

Figure 14-26   The fixed probe is a plunger-operated proximity switch that can be permanently mounted on the machine and connected by cable to the MCU. (Courtesy of Cincinnati Milacron)

checking and setting before use. If a redundant copy is not found, a feed-hold on the machine cycle is activated, the FMS computer is notified, and the part program branches to the program end or a safe area within the same program. Jumping to the program end will result in a new workpiece moving into the work queue for processing (the incomplete workpiece moves to a review stand), while jumping to a safe but different area in the same part program permits machining to continue on the same workpiece.

As with tool monitoring, part program alignment must be maintained in a feed-hold or stop condition. And broken tool sensing, like torque measurement, is controlled through conditional branching and statement labels within the NC part program.

CUTTING TOOL CONSIDERATIONS AND GUIDELINES

Because cutting tools are up front and actually performing the metal-removal work (all the other FMS hardware and software elements are just support), it is important to discuss some general cutting tool factors and related guidelines.

A stand-alone NC machine can be expected to cut metal only about 20 percent of the time. A CNC machining center with dual pallets, as seen in Figure 14-27, will increase actual cut time to around 75 percent, while the same CNC machines operating within an FMS can be expected to be cutting metal 90 to 95 percent of the time. Although cutting tools in an FMS are no different and cut metal no faster than those used for stand-alone applications, such increases in productiv-

Figure 14-27 A CNC machine with dual pallets can be expected to be "in the cut" approximately 75 percent of the time. (Courtesy of Kearney and Trecker Corp.)

ity rates result in more tool use in a given period of time. Consequently, cutting tools are used up 4.5 times as fast in an FMS application. It becomes important, therefore, to recognize that, when buying cutting tools, cutting tool life, not cutting tools themselves, is the real purchase. Decisions for buying cutting tools should receive the same close scrutiny considerations as other FMS components.

Regardless of the type of cutting tool used (drills, taps, end mills, reamers, lathe tools, and so on), two variables serve as simple criteria by which to judge the machinability of various metals. These are chip formation and cutting fluid. Chip formation is important as a means of determining the degree of finish the workpiece will have, in addition to revealing the efficiency of the machining operation. The proper type of cutting fluid will provide better accuracy and efficiency of operation.

Chips formed in the machining of metal, which vary in type and in their effect on surface finish and tool wear, fall into three classes:

1. Discontinuous
2. Continuous
3. Continuous with edge buildup

The discontinuous type of chip, as shown in Figure 14-28, is formed when the metal that is forced upward over the tool face is broken into short segments. This

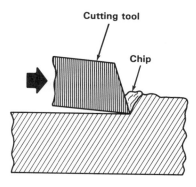

Figure 14-28   A discontinuous chip is formed when machining brittle metals.

type of chip is caused by brittleness in the metal being machined and is present in the machining of cast iron and other brittle metals.

The continuous chip (Figure 14-29) does not produce a built-up edge on the tool face. It is formed by continuous deformation of the metal being machined, the metal ahead of the tool deforms without fracture, producing a chip that travels smoothly up the tool face. As in the case of discontinuous chip formation, the face of the tool becomes worn by the abrasive action of the chip sliding over it so that the cutting edge is slowly rounded and worn away.

The continuous chip with edge buildup (Figure 14-30) forms on metals that have good ductility (easily formed, molded, or shaped). A compressed mass of metal sticks to the face of the cutting tool. Portions of the built-up edge break off from time to time and are carried away by the chip. Other portions of the built-up edge stick to the workpiece and cause roughening of the machined surface. This type of chip will cause poor surface finish and severe wear on the cutting tool.

Coolant use in FMS or stand-alone NC machine applications (Figure 14-31) is an important consideration to cutting tool life, accuracy, and efficiency because it:

1. Cools and carries heat from the cutting edge of the tool and workpiece
2. Lubricates and reduces heat friction at the tool face
3. Allows smooth chip flow off the cutting edge

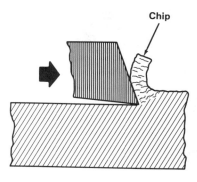

Figure 14-29   A continuous chip is formed when machining ductile metals.

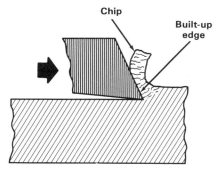

Figure 14-30   A continuous chip is formed when machining ductile wear-resistant metals.

4. Backflushes chips
5. Reduces tool wear and power consumption
6. Improves surface finish
7. Dampens flying dust or fine particles that are a safety hazard to operating personnel

Many tools used in both FMS and stand-alone NC machine applications are of the high technology or indexable insert type as seen in Figure 14-32. Such tools were designed to replace conventional cutting tools because they can run, depending on material type, up to ten times faster at coated insert speeds and feeds. Use of these indexable insert tools reduces delay time because only the inserts need to be replaced, not the entire tool. Also, inserts have two, three, four, or more cutting

Figure 14-31   Coolant use is an important consideration to cutting tool life, accuracy, and efficiency. (Courtesy of Cincinnati Milacron)

edges per insert, thereby permitting indexing to a new cutting edge on the same insert before throwing away and replacing the entire insert.

Another important factor relative to use of indexable insert tooling is that it does not require regaging for tool length each time an insert cutting edge needs replacement. Tool and insert tolerances are close enough to avoid this lengthy and sometimes error-prone procedure.

Whether indexable insert tooling is used on machining center or turning center applications, heat is still the primary cause of insert wear and destroyer of insert life. Shearing action of the metal being formed into a chip (Figure 14-33) and the chip rubbing against the cutting tool are sources of the heat. Other sources of insert wear include excessive or inadequate feeds, speeds, and depth of cut and various forms of tool deformation.

The most common forms of insert deformation causing wear are:

☐ Cratering
☐ Chipping

Cratering (Figure 14-34) is caused by improper insert selection based on the alloy content of tungsten carbide with titanium and tantulum carbides. The improper alloy content causes a lack of resistance to crater and flank wear, resulting in cutting-edge deterioration and breakdown.

Chipping (Figure 14-35 on page 317) is caused by impact to the insert cutting edge or vibration and lack of rigidity in the machining setup. Using an insert cutting edge that is not crack or vibration resistant or too hard a grade of carbide for a particular application promotes chipping. Material may weld itself to the insert cutting edge if cutting speeds are too low. When the built-up material is removed, a portion of the cutting edge may also be removed, thereby causing a small chip in the insert.

Any discussion of cutting tool considerations should include tool holders (Figure 14-36 on page 317). Spindle-type tool holders essentially hold and drive the

Figure 14-32   Indexable insert cutting tools are replacing conventional cutting tools because they can run faster and have higher penetration rates, and only the inserts need to be replaced. (Courtesy of Kennametal, Inc.)

Figure 14-33   Shearing action of metal being formed into a chip and the chip rubbing against the cutting tool is the heat source in chip formation and flow. (Courtesy of Cincinnati Milacron)

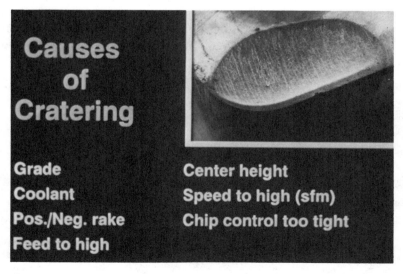

Figure 14-34   Example of cratering, resulting in cutting-edge deterioration and breakdown. (Courtesy of Kennametal, Inc.)

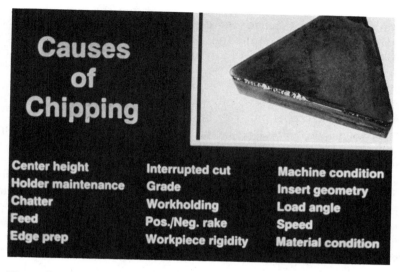

Figure 14-35 Example of chipping caused by impact to the cutting edge, vibration, or lack of setup rigidity. (Courtesy of Kennametal, Inc.)

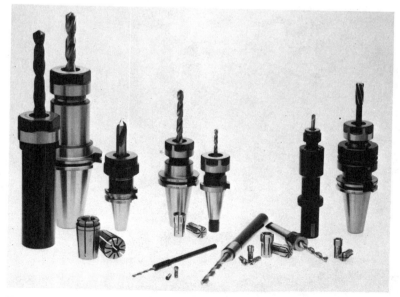

Figure 14-36 Examples of spindle-type tool holders and accessories. (Courtesy of Kennametal, Inc.)

entire assembly, as seen in Figure 14-37, which is made up of the basic cutting tool, adapters or collets, and the tool holder itself. Most tool holders conform to the ANSI industry standard for tapered V-flange tool holders. Dimensions are critical on flange, shank, and retention studs. Conformance to such rigorous tolerances and ANSI standards is mainly for safety and interchangeability requirements. The ANSI standard covers six basic sizes of toolholders from # 30 through # 60. Each machine tool accepts only one basic size; # 50 is commonly used on horizontal machining centers that have from 10 to 40 horsepower. The most important aspect about tool holders is that they be free from nicks, gouges, dirt, grit, and any other visible signs of damage. Lack of tool-holder cleanliness can cause the entire tool assembly to "run out," resulting in part inaccuracies. Tool holders must be cleaned, inspected, reconditioned, if possible, or scrapped if accuracy or safety becomes questionable. Tool holders must accurately drive the cutting tools and run true. Also, machine tool spindles and tool matrix pockets should be periodically inspected and wiped out (while stopped and in the manual mode) to remove dirt, chips, and grit. Gouging of the machine spindle or the tool holder can result when pull-back clamping pressure is applied. Accepting anything short of an optimal tool assembly is settling for second-rate performance of the entire FMS.

Turning center tool holders using disposable inserts, as seen in Figure 14-38, permit indexing a carbide insert that has six to eight cutting edges in less than 1 or 2

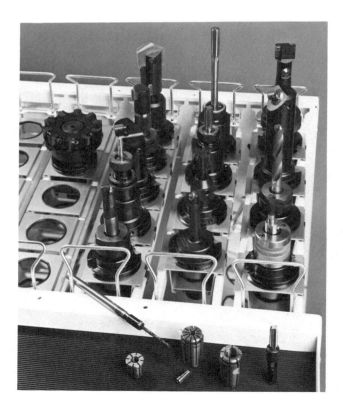

Figure 14-37   Spindle-type tool holders hold and drive the entire tool assembly. (Courtesy of Kennametal, Inc.)

Figure 14-38   Turning center tool holders using disposable inserts permit indexing a carbide insert that has six to eight cutting edges in less than 1 or 2 minutes. (Courtesy of Kennametal, Inc.)

minutes. If part tolerances are relatively open, the original tool setting can be maintained, and the machine is ready for production with no additional tool adjustment. Generally, as long as the same holder is used on an operation, the tolerances within the holder will have little effect on the ability of the insert to index to close tolerances.

Regardless of whether CNC machines are used in stand-alone, cell, or FMS applications, sound cutting tool guidelines and qualified practices must be followed if any machine is to perform to its full capacity. Each machining application and situation requires its own careful analysis and study of conditions as there are no general solutions to solve all potential tooling problems. However, some general cutting tool guidelines and practices should be followed:

1. Always choose the shortest drill length that will permit drilling any hole to the desired depth.
2. Check and maintain correct cutting feeds and speeds for all tools. Optimum feeds and speeds may not always be achievable, depending on machinability and material characteristics, rigidity of setup, and so on.
3. Maintain rigid and tightly enforced standards with respect to tool setting, regrinding, and tool compensation.

4. Tool assembly dimensional standards must be rigidly enforced. Part programmers must be given cutting tool standards to count on.

5. All tools must be checked before they are used. This includes the cutting edges and tool body, as well as holders, extensions, and collects. All should be in perfect order and able to act as a total tool assembly.

6. Select the right tool for the job and use the tool correctly. Tools should be able to machine the workpiece to the desired accuracy.

7. When purchasing cutting tools and tool holders, examine the cost per piece part produced as well as the cost of the tooling package. In many cases, bargain tools cost more per part produced.

8. Excessive cutting speeds generate excessive heat, resulting in shorter cutter life. Chip and cutter tooth discoloration are good indicators of excessive cutting speeds.

9. Use coolants to get maximum cutter life and to permit operating at higher cutting speeds.

10. Understand machine and control capabilities. Additional tools are often purchased because built-in machine and control capabilities such as contouring are not used effectively.

11. Care for tools properly. This includes perishable tools, holders, drivers, collets, inserts, extensions, and others. They represent a sizable investment and should be adequately stored and reconditioned when required.

12. Watch and listen for abnormal cutting tool performances. Attention paid to actual metal-removal processes can often prevent tool breakage problems, scrapped parts, and rework. Chatter and other vibration abnormalities can be corrected if detected early enough.

13. Select standard tools whenever operating conditions allow. Standard tools are less expensive, readily available, and interchangeable.

14. Check and make sure tool holder tapers, spindle tapers, and tool holder insert pockets remain clean and free of chips, grit, and dirt.

SUMMARY

1. FMS scheduling of workpieces through the system is driven by tool capacity.

2. Tool management is getting the right tool to the right place at the right time.

3. The ability to sense and respond to tool fault conditions is an important issue of FMS tool management.

4. Although tool management is one of the most difficult aspects of FMS to regulate and control, it is also one of the most vital.

5. The four required elements of tool management systems are tool room service, tool delivery, tool allocation and data flow, and fault sensing.

6. The most accepted and used tool strategies are mass exchange, tool sharing, tool migration, and assigned tools.

7. A primary concern of automated tool exchange is to eliminate spindle interruption.

8. Automated identification systems are important to FMS applications because they are reliable, save time, and reduce human error.

9. Bar code scanning is the most widely used form of automated identification.

10. Tool monitoring compares a tool's actual cutting time against its predetermined cutting life.

11. Torque measurement and control are essential to FMS because of their ability to adapt feed rates under varying stock-removal conditions as well as to sense broken or dull tools during unattended operation.

12. Tool fault detection alone without a versatile and automatic recovery strategy offers little advantage in an unattended system.

13. Gathering and compiling machinability information for a tool life data base are the responsibilities of each user.

14. The ability to automatically and programatically gage, check, and set tool lengths is a requirement in an FMS.

15. Chip formation and cutting fluid are used to judge the machinability of various metals.

16. Chips formed in the machining of metal fall into three classes: discontinuous, continuous, and continuous with edge buildup.

17. The most common forms of insert deformation causing wear are cratering and chipping.

REVIEW QUESTIONS

1. To make intelligent cutting tool quantity–cost buy decisions, it is of little value to closely monitor and review actual consumption. True or False?
2. Tool preset personnel can make "free substitutions" for tools that are unavailable but called for in the part program True or False?
3. The flexibility of an FMS is constrained by two support resources, palletized workpieces and _____.
4. The basic concept of tool management is:
    (a) Maintaining sufficient redundant tool back-up at the machine
    (b) Automating the tool preset area
    (c) Getting the right tool to the right place at the right time
    (d) Monitoring, detecting faults, and replacing tools as required
5. List and describe the principal elements of tool room service.

6. The most important element of an automated tool delivery and exchange system is to minimize spindle interruption. True or False?

7. Tool allocation and flow are:
   (a) Assigning and controlling the number of tools required for each machine to process the FMS parts
   (b) Knowing which tools to use in proper sequence
   (c) A method of automated tool delivery
   (d) Based on infinite tool capacity

8. _____ refers to the time during which a cutting tool produces acceptable parts in a machining operation.

9. The most accepted and used tool strategies are tool sharing, tool migration, assigned tools, and _____ .

10. The assigned tool strategy identifies the most-used tools for the production requirements and assigns permanent residence to those tools in each machine tool matrix for the full production run. True or False?

11. The mass-exchange tool strategy permits tool-exchange control to be minimized at the expense of an increase in tool inventory. True or False?

12. _____ are made up of binary digits arranged so that the bars and spaces in different configurations represent numbers, letters, or other symbols.

13. Name four identification systems available and being used in FMS and other factory automation applications.

14. Tool monitoring measures normal tool wear against a predetermined standard stored on each machine's MCU True or False?

15. Unpredicted catastrophic tool failures can be detected by:
    (a) Checking each tool's known set length before use
    (b) Measuring a tool's torque at the spindle or through tool fault and sensing detection
    (c) Scanning the tool matrix for any broken tools
    (d) Only (a) and (b)
    (e) All of the above

16. Look-ahead capabilities exist in FMS software that can tell when cutting time will expire on a given tool and determine how many duplicates are required. True or False?

17. Broken tool sensing and _____ are controlled through conditional branching and statement labels within the NC part program.

18. Chip formation and cutting speed are used to judge the machinability of various metals. True or False?

19. Heat is the primary cause of insert wear and destroyer of insert life. True or False?

20. List four reasons why coolant use is an important consideration to cutting tool life, accuracy, and efficiency.

21. _____ is caused by improper insert selection, which causes cutting edge deterioration and breakdown.

22. _____ is caused by impact to the insert cutting edge, vibration, or a lack of rigidity in the machining setup.

15

Workholding Considerations

OBJECTIVES

After studying this unit, you will be able to:

☐ Understand the importance of fixturing to a successful FMS installation.

☐ List and discuss some basic fixture support and location principles.

☐ Contrast the differences between hard versus modular fixturing.

☐ Describe some general fixture considerations in an FMS environment.

INTRODUCTION

An FMS is only as successful as the amount of detail designed and built into each component or element of the system. Fixtures are one of those important elements.

Poorly designed fixtures, sloppy manufacturing methods, and "make it fit" fixture build and assembly practices are problems, even in a stand-alone NC environment. Fixture problems hold up production, cause scrap and rework, interfere with delivery schedules, and cause extra delays through finger pointing and blame placing among operating personnel.

Fixtures take on an increasingly important role in an FMS environment because of the increased machine tool utilization rates and round-the-clock operations.

In an FMS application, just as with stand-alone NC machines, the basic fixture requirements are still that parts be accurately positioned, located, and clamped in position. However, in an FMS environment, fixtures must be designed to accommodate part families with tighter tolerances and broader ranges in size, while minimizing changeover time. Consequently, fixtures designed and built for FMS applications demand acute attention to detail and an understanding and awareness of the environment in which they will be used.

GENERAL FIXTURING GUIDELINES

Proper fixturing is extremely important to an NC installation whether in a stand-alone, cell, or FMS environment. Poorly designed or manufactured fixtures can cause numerous problems that can readily be seen in terms of scrapped parts, fixture rework costs, and extensive and time-consuming delays. If proper design considerations are applied, these costly and extensive delays can be avoided.

The primary function of any fixture (Figure 15-1) is to locate and secure the workpiece for succeeding machining operations. This involves initial setup time for the workpiece to be loaded and clamped in the fixture for machining. Loading and unloading constitutes an important part of the nonproductive cycle time of each part. By simplifying this operation, more parts can be produced per hour. Therefore, fixtures should be designed to reduce part setup time.

NC machines normally permit smaller batch lots, thereby reducing part inventories. As a result, fixtures will be set up on a stand-alone machine and used more frequently. This will warrant much consideration of design to reduce part

Figure 15-1   Typical work-holding fixture. (Courtesy of Mid-State Machine Products)

setup time and simplify the process of securing the fixture, as danger exists for both overdesign and underdesign in these areas.

Part accuracies requiring work-holding fixtures depend on the accuracies of the fixture itself regardless of how well the part is programmed or processed. Money spent taking the time to cover all aspects of locating and holding the workpiece accurately and securely will pay dividends when part and fixture reach the shop floor. This time, money, and effort should be spent up front in the initial design phase before production begins.

Fixtures must hold workpieces accurately and rigidly while providing, in some cases, multisided part access for all cutting tools, including their approach and departure paths. This means a minimum of protruding clamps and fixture detail components. Clamps should always be placed as close to the support locations as possible (Figure 15-2). Placing unsupported clamps at any convenient location on the fixture could mean distortion of the part under clamping pressure. Clamp positions for ease of loading and unloading and for cutting tool accessibility should always be given a priority. Clamps must avoid blocking hole locations and milling cuts that may interfere with part processing. Tool holders should also be considered, particularly when the cutter is engaged in the workpiece, so as not to interfere with clamp assemblies and other fixture details during machining.

Clamping should become increasingly automated with larger part volumes (Figure 15-3). Types of clamps that could be used are pneumatic- and hydraulic-actuated clamps, toggle clamps, and cam- actuated clamps. Robotized clamping can be considered in some cases, depending on the application. Regardless of the type of fixture to be used, the application, or the number of parts to be processed,

Figure 15-2   A fixture with part clamped in position showing close placement of clamps to support locations. (Courtesy of Cincinnati Milacron)

Figure 15-3 Holding fixture with automated clamping. (Courtesy of Jergens Power Clamping)

fixtures should always be designed with safety in mind. Operators should always be able to reach easily all clamps and adjusting screws that may need attention. Sharp corners should be minimized, and access for chip removal should be easy. The best designed fixture is also the safest for those who must use it.

Part orientation is an important consideration in fixture design. Although sometimes overlooked, fixtures should always be designed to prevent incorrect loading and part orientation. Foolproofing of fixture design to prevent incorrect part loading and orientation is critical; mistakes can easily happen when work is highly repetitive. To foolproof a fixture design, parts should be able to be located and clamped only one way. This may take additional time to develop, but it will help reduce scrapped parts, broken tools, embarrassing mistakes, and possible injury.

The advantages of sound economical fixture design, manufacturing, and assembly are:

1. Reduced fixture-to-machine and part-to-fixture setup time
2. Decreased cost per part
3. Reduced fixture and part inspection time
4. Better consistency of part accuracies
5. Minimal fixture modifications or rework
6. Faster and easier NC program prove-out where required
7. Reduced errors and inaccuracies in part location and clamping

Many aspects must be considered regarding the design and application of fixtures. Some of these include:

1. Fixtures should be fitted with precisely located tram buttons at the part's reference point to which the operator can easily touch up in precisely locating the fixture for alignment purposes.

2. Maximum and minimum tool lengths should be considered to determine optimum positioning of the fixture-to-table and part-to-fixture relationships. All cutting tools must reach and be able to easily access surfaces to be machined without having to add tool extensions and special adaptors.

3. The fixture-to-machine table mounting orientation must be defined. Much time and effort are often spent designing a foolproof part-to-fixture relationship and the fixture is loaded and clamped 90 or 180 degrees out of position. This can have expensive and hazardous consequences if the fixture is not marked to precisely orient it to the pallet or machine table.

4. The fixture should be designed and placed on the machine table to support efficient part processing. In some cases, inefficient NC programming moves are required to avoid poorly designed and placed locating pads and clamp assemblies. Fixtures designed for horizontal machining centers should ideally be positioned over the center of rotation. This should provide an equal distance from the center of rotation to each part face to be machined. It should also promote programming simplicity in X, Y, Z and B axes.

5. Fixture design should be simple, and standard components should be used whenever possible.

6. The fixture and part should be located positively. Rough, nonflat parts should be supported in three places and located on tooling holes, if possible. Fixture bottoms and locating keys, along with machine table or pallet locating surfaces, must be clean and and free of dirt, burrs, chips, and surface irregularities prior to fixture mounting.

7. Fixture component parts should be easily accessible and movable during any part of the load or unload sequence and replaceable in exactly the same position.

FIXTURE SUPPORT AND LOCATION PRINCIPLES

Tool design is the process of designing and developing the tools, methods, and techniques necessary to improve manufacturing efficiency and productivity. The main objective of any tool or fixture design is to lower manufacturing costs while improving quality and increasing production.

The pressures of modern industry place increasing demands on tool and fixture designers, particularly in the areas of maximizing productivity at minimal cost and improving part accuracies through design simplicity and use of standard components. Every detail should be considered for possible savings in time and material. Overly elaborate designs only serve to increase costs without adding significantly to accuracy or quality. All fixtures should be made as simple as the part design permits.

When designing special tools and fixtures (Figure 15-4), the designer must keep the part tolerance in mind. As a general rule, the tool tolerance should be

Figure 15-4 Tool designers must keep part tolerances in mind when designing fixtures. (Courtesy of Cincinnati Milacron)

between 20 and 50 percent of the part tolerance. This is necessary to maintain the required precision. Specifying tool tolerances closer than 20 percent only increases the cost of the tool and adds little to workpiece quality. Generally, tolerances greater than 50 percent do not guarantee the desired precision. Accuracy of the part being machined is the single determining factor as to what the tool tolerance should be.

    Component parts are made in almost every possible shape and size. The tool designer must be able to accurately locate each part, regardless of how it is made. To do this, the tool designer must know the various types of locators and how each should be used to get the best workpiece placement with the least number of locators.

    There are three primary methods of locating work from a flat surface (Figure 15-5): solid supports, adjustable supports, and equalizing supports. These types of locators set the vertical position of the part, support the part, and prevent deflection and distortion during part processing.

    Solid supports are the easiest to use. They can either be machined into the tool base or installed. This type of support is normally used when a machined surface acts as a locating point or position.

    Adjustable supports are used when the surface is rough or uneven, as in cast parts. There are many types and styles of adjustable supports. A few of the more common styles are the threaded, spring, and pusher type. The threaded type is the easiest and most economical and has a larger adjustment range than the others. Adjustable locators are normally used with one or more solid locators, thereby promoting the adjustment needed to level the workpiece.

Figure 15-5   A fixture showing part location from a flat surface with solid supports (locators). (Courtesy of Cincinnati Milacron)

Equalizing supports are also a form of adjustable support. They provide equal support through two connected contact points. As one point is depressed, the other raises and maintains part contact. This type of support is especially necessary on uneven cast surfaces.

The terms locators and supports are used interchangeably when discussing the devices used under a workpiece. Locating devices used to reference the edges of a part are generally called locators or stops.

When choosing a support, the tool designer must consider the shape and surface of the part and the type of clamping device to be used. The support selected must be strong enough to resist both the clamping pressure and the cutting forces.

Locational inaccuracies develop because of the difference in position and location tolerances between the fixture and the workpiece. For example, problems can be created by locating a part from both its outside edge and some previously machined holes. These problems originate from the fact that the fixture pin locations are fixed and cannot be changed to suit each part. Workpiece hole locations can be expected to vary within the print dimensional tolerances, and when a part is placed in the fixture that is at either extreme end of the part tolerance, it may not fit. Hole locators can obviously be made smaller to accommodate part variation, but this renders the hole locator useless and defeats its purpose. To avoid this problem, tool designers should never attempt to locate a part from both its previously machined holes or edges. Only one should be specified and used.

In the initial stages of machining, it is most common to locate workpieces from an external profile or an outside edge. Profile locators position the workpiece in relation to an outside edge or an external detail such as a hub or boss. Other types of initial machining fixture locators include fixed-stop, adjustable stop, dowel pins, and V locators. Tooling holes (holes used for location for succeeding machining operations), are usually machined in the initial stage of part processing.

Locating a part from a premachined hole is the most effective way to accurately position the workpiece. Out of twelve possible directions of movement, nine are restricted by use of a single location pin, and eleven are restricted with two locating pins. If possible, tooling holes should be premachined in each part and used as primary part locators for succeeding machining operations.

The primary difference between pins used for location and pins used for alignment is the amount of bearing surface. Alignment pins usually have a longer contact area. Locating pins usually have a contact area of one-eighth to one-half of the part thickness. More than this makes workpiece loading and unloading a difficult operation.

Another style of locating pin common to jigs and fixtures is the diamond or relieved type. This pin is normally used along with the common round type (Figure 15-6) to reduce the time required to load and unload each workpiece in the fixture. It is easier to locate a part on one round and one diamond pin than to locate on two round pins. This is because use of a diamond pin permits some ease of movement during part loading and unloading. However, with the part located in the holding fixture, the round pin locates the part while the diamond pin prevents movement around the pin. To be effective, the diamond part of the pin must always be placed to resist this movement.

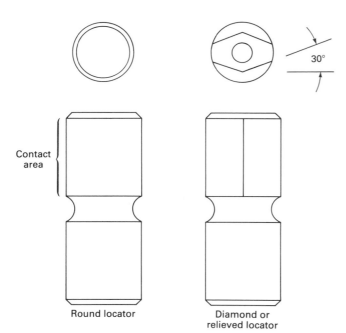

Figure 15-6   Round and diamond (relieved) locators are used to locate parts on premachined holes.

Fixture location also involves locating the fixture to the table or pallet. Fixtures must be designed for use on certain machines because of different locating and clamping provisions. Some permit fast fixture-to-table interchange by utilizing edge locators mounted to the machine table, usually two at right angles to each other. The fixture is placed on the table and pulled back into the corner formed by the two blocks and then clamped down. However, edge locators tend to become misplaced because excessive forces or dirt interferes with snugging the fixture all the way back, or poor clamping pulls the fixture away from the positive seating of the locators.

Some rotary index tables are frequently fitted with a plug precisely located at the center of index. When mounting a fixture, a button or pin built into the center of the fixture is dropped into the table's plug. The fixture is then pivoted until a tongue drops into a precision keyway in the table to provide precise angular orientation and location.

Multiple tongues that run 90 degrees to each other are used in some cases. However, on a rotary index table (Figure 15-7), a slight tongue slot clearance on one side might double the error when the table is indexed 180 degrees. This could

Figure 15-7   Fixtures can be oriented to rotary index tables by means of tongues or side locators. (Courtesy of Kearney and Trecker Corp.)

potentially be a problem if critical bores, 180 degrees apart, are required to line up. However, for most horizontal machining center work, precision tongues in machine table or pallet tongue slots are adequate fixture locators.

MODULAR FIXTURING

The concept of modular fixturing, gaining wide acceptance among manufacturers, is not new. Fixturing systems based on a family of interconnecting components have been used in the United States since the early 1940s. Over the past several years, however, interest in modular fixturing has been increasing due primarily to manufacturers striving to reduce costs in the face of intensified worldwide competition.

Modular fixturing (Figures 15-8 and 15-9) employs the use of reusable, interchangeable components that can be assembled and torn down based on changing part requirements. The objectives behind modular fixturing are the same as those for table setups or dedicated fixtures; the part must be positioned, located, and held securely in place. Modular fixtures rely on a base plate or subplate (Figure 15-10) or a tombstone-type base (Figure 15-11 on page 334) for the foundation of the fixture. Both base plates and tombstone-type bases can be mounted directly to the machine table in a stand-alone NC machine environment or to a pallet in a cellular or FMS environment.

In addition to foundation elements, modular fixturing permits the selection of precision modular components (Figure 15-12 on page 334) manufactured to tolerances as close as 0.0002 to 0.0004 inch (0.005 to 0.01 mm) that serve as building

Figure 15-8   Modular fixture assembled with workpiece clamped in position. (Courtesy of Cincinnati Milacron)

Figure 15-9 Another modular fixture with workpiece clamped in position. (Courtesy of Cincinnati Milacron)

Figure 15-10 Part located and clamped on a modular fixture base or subplate that mounts to a pallet or machine table. (Courtesy of Qu-Co Modular Fixturing, a TE-CO Co.)

Figure 15-11   Modular fixture tombstone base used for clamping parts for horizontal machining center part processing. (Courtesy of Qu-Co Modular Fixturing, a TE-CO Co.)

blocks for constructing fixtures. This usually includes a range of single-point positioners, single and multiple surface locators, and clamping devices. Generally, the same style of clamps used for table setups or dedicated fixtures is used with modular fixtures. These include plain, step, round nose, U shape, toggle, edge, toe, and swing clamps. Ancillary components exist as well, which fall into logical groups such as adjustable stops, adapter plates, subplate mountable vises, and an assortment of alignment pins, studs, bolts, nuts, and washers.

Modular fixturing affords users the following advantages:

1. Fixtures can be assembled and torn down and components reused.
2. Dedicated fixture design, component manufacturing, inspection, and rework costs can be substantially reduced or eliminated.
3. Elimination of formal fixture design and manufacturing saves valuable lead

Figure 15-12   Modular fixturing precision components can be assembled much like an erector set. (Courtesy of Qu-Co Modular Fixturing, a TE-CO Co.)

time and reduces coordination problems relative to fixture design, manufacturing, and assembly.

4. Dedicated or hard fixtures (those dedicated to only one part or part family) are scrapped when part life becomes inactive; modular components can be salvaged and reused.

5. If engineering change necessitates alteration, hard fixturing requires lead time, rework, and inspection; modular components can easily be moved or replaced with other modular components as required.

6. Hard or dedicated fixturing requires extensive storage facilities; modular components can be stored in bins (Figure 15-13) and assembled as needed.

7. The fixture designer is never locked into his or her design since alternative fixturing solutions can be tried easily by relocating clamps, stops, and locators.

8. The part-to-machine setup and orientation time required for NC jobs using table setups (conventional nondedicated work-holding componentry) can be reduced or eliminated; modular fixtures can be accurately re-created and assembled off-line in a fixture setup area in advance of machining operations.

Modular fixtures can be designed by trial and error without a formal design using a conventional fixture design approach. A major advantage of modular fixturing is that the fixtures can be constructed at the last minute, once the part to be machined is available. Many times better fixtures are built by trial and error since the problems associated with part orientation can be readily visualized. It is not uncommon for new modular fixtures to be completely built from scratch in less

Figure 15-13   Modular fixturing components can be neatly stored in bins and assembled as needed. (Courtesy of Cincinnati Milacron)

time than it takes to prepare the cutting tools for the job. And reconstructing a modular fixture that has been built once and had its design documented requires about a quarter of the time.

Companies with CAD systems can create a library of fixture components on their systems to further reduce design time, especially if the part to be fixtured is already represented in the CAD data base. Once a library of fixture components has been created on a CAD system, components never need to be re-created. A modular fixture is designed on a CAD system by recalling the desired fixture components from the library and placing them, where needed, until the fixture is complete. Utilizing this CAD fixture design methodology, several different fixturing solutions can be tried via the CAD screen for the same part to determine the best design without actually having to build each fixture. Since CAD systems are inherently accurate, typically in excess of five decimal places, the fixture designer has the added advantage of being able to check for fixture component–workpiece interferences.

FIXTURE CONSIDERATIONS IN AN FMS ENVIRONMENT

Fixture applications in an FMS environment require additional considerations over and above standard or conventional requirements and guidelines. These additional considerations exist because of the round-the-clock and lightly staffed operating mode of FMS, along with the tremendous cost involved in purchasing the system. Such considerations are generally true of fixturing in any automated environment. However, in an FMS environment, system flexibility, based on the number of dissimilar parts required to run through the system, places additional demands and constraints on fixture design.

The basic strategy when designing fixtures for FMS application (Figure 15-14) is to complete as many machining operations as possible in one clamping of the workpiece. This may not be possible in all cases, as tooling holes may need to be machined in the initial pass and the part may require another clamping and an additional pass through the system for completion.

Some primary FMS fixture considerations would be:

1. Pressure to complete as many of the machining operations as possible in one pass places the tool designer under extremely tight constraints. It limits clamp placement availabilities and, in some cases, depending on part design and attributes, may actually take longer to machine. This could be true because limited clamping opportunities may necessitate nonrigid clamping, thereby minimizing aggressive machining. Consequently, more milling cuts, for example, might be required to complete machining.

2. In an FMS, loading the fixture is done internal to the FMS cycle (can be done while the system and machine tools are in cycle) at a load–unload station (Figure 15-15 on page 338). However, fixtures must be designed to minimize part handling time because the FMS is lightly staffed, and some part cycle times may be short. Consequently, operators may have difficulty accurately locating the parts

Figure 15-14   Typical fixture on a pallet in an FMS application. (Courtesy of LTV Aircraft Products Group)

and keeping up. Also, some fixtured parts may be rather large, heavy, and unwieldy. To load and unload parts of this type in a vertical fixtured position, even with crane support, would be extremely difficult and hazardous. As a result, a tilt table (Figure 15-16 on page 339) may be required to bring the part and fixture from a vertical to a horizontal position to facilitate loading and unloading.

3. Fixtures designed to be used for FMS applications require the designer to be knowledgeable of machining principles and general practices in order to provide a design that contributes to using a minimal number of cutting tools to machine the part. Cutting tool assemblies are expensive, take time to assemble, disassemble, maintain, change, and transport, and use up tool pockets in a machining center's tool matrix. Therefore, fixtures should be designed to minimize tool use and to promote ease of cutting tool-to-part accessibility. For example, use of an end mill to profile mill a cored opening in a casting through circular interpolation can eliminate multiple boring bars, provided the fixture has been designed to permit ease of milling cut accessibility.

4. Operating personnel responsible for loading and unloading parts must be well acquainted with and exercise the principles of proper part location. Lack of proper part location, which includes cleanliness, checking for solid location before and after clamping, and distortion under clamping pressure, all can contribute to

Figure 15-15   Fixtures in an FMS at a load–unload station. (Courtesy of Kearney and Trecker Corp.)

workpiece scrap and rework. In some cases, specific loading instructions may be required, depending on part dimensional accuracies, fixture and/or part complexity, setup rigidity, and other factors. Such instructions could be a step by step itemized listing, for example, of how the part should be loaded and located and which clamps to tighten in which order. This could also include use of a torque wrench to tighten bolts and nuts to exact clamping pressures.

5. In an FMS environment, fixtured pallets are transported to and from processing stations via AGVs (automated guided vehicles), as seen in Figure 15-17. Although pallets are designed and manufactured to be interchangeable and accept any fixture, cumulative error buildup may exist among part-to-fixture, fixture-to-pallet, and pallet-to-machining center, thereby possibly machining some parts out of tolerance. Consequently, some fixtures, depending on required tolerances, must be "married" or pinned to a particular pallet to reduce the interchangeable error buildup problem.

6. Depending on the number of different fixtures used in an FMS and their frequency of use, fixture storage, retrieval, and identification can become a problem. A detailed analysis of part scheduling requirements and frequencies, along with a review of FMS floor space allocation, layout, and accessibility for fixture storage and retrieval, will provide insight as to how these problems might be avoided.

Figure 15-16 Tilt table about to bring a tombstone from a vertical to a horizontal position for ease of part loading. (Courtesy of Cincinnati Milacron)

Figure 15-17 Fixture and workpiece being moved from one processing station to another in an FMS. (Courtesy of General Dynamics Corp., Fort Worth Division)

7. Sometimes problems originally thought to be fixture related are caused by other sources. A potential source of problems in the case of castings might be excessive or insufficient stock allowances, for example. Such inconsistencies can point to the fixture and improper part location as the problem source because parts drift in and out of required tolerance. However, the problem cause may lie with the casting vendor's inability to qualify and control the process. In an FMS environment, casting vendors must consistently meet stringent raw material guidelines and qualification criteria in order to reduce or eliminate problems of this type.

8. Due to potential cumulative error problems, parts and fixtures in lightly attended machining environments like FMS require probing (Figure 15-18). Probing for excess or insufficient workpiece stock allowances and accurate fixture alignment permits automatic machine reorientation to suit changed conditions in X, Y, or Z as detected by the surface-sensing touch probe. Such part, fixture, and alignment conditional checking, however, can add approximately 50 percent more time to program each part because of the additional conditional and branching logic required in the part program.

9. In many cases, parts fixtured for machining in an FMS remain in that same fixture for CMM (coordinate measuring machine) inspection. Consequently, parts inspected in the holding fixture may check in tolerance, but when inspected out of the holding fixture in a free state, they may check out of tolerance. This happens because the part springs back when clamps are loosened. In many cases this is brought about by poor placement of fixture locators or clamps or inconsistent or inadequate clamping pressures.

Although some of the previously mentioned fixture considerations apply to

Figure 15-18   Probing is used extensively in an FMS to check for part presence, stock allowances, and accurate fixture alignment. (Courtesy of General Dynamics Corp., Fort Worth Division)

stand-alone and cellular NC machine applications, the overall tool design and fixturing effort in an FMS requires considerable more thought, skill, attention to detail, and general awareness.

Fixture building (Figure 15-19) in an FMS is generally done at a fixture build station, which aids the operator in performing the following operations:

1. Location and retrieval of fixture components
2. Building of pallet and fixture assemblies from components for a work order
3. Adding and removing fixtured pallets from the system
4. Disassembly of pallet and fixture components when workpiece processing is complete
5. Storage of fixture components or assemblies

At a scheduled time, prior to authorization of a work order, the system manager will download fixture and pallet assembly build instructions. The operator's function at the fixture build station is to obtain the required fixture com-

Figure 15-19   Fixture building in an FMS is generally done at a fixture build station. (Courtesy of Kearney and Trecker Corp.)

ponents from the component storage area and build a complete fixture and pallet assembly. Fixtures must be mounted on pallets to be transported through the system. Once complete, the fixture and pallet assembly is then labeled and identified to the system. A system entry–exit position for the entire fixture and pallet assembly is then determined, and the assembly is manually transferred to that position. Operator input to the station control console then causes the pallet identification to be read to verify the operator's action. After verification of the pallet ID, the FMS computer is informed of the availability of the fixture and pallet assembly for manufacturing duty. A flow chart, depicted in Figure 15-20, illustrates the sequence of this process. When the fixture and pallet assembly is no longer required, it will be delivered to the entry–exit position for removal from service when an operator becomes available.

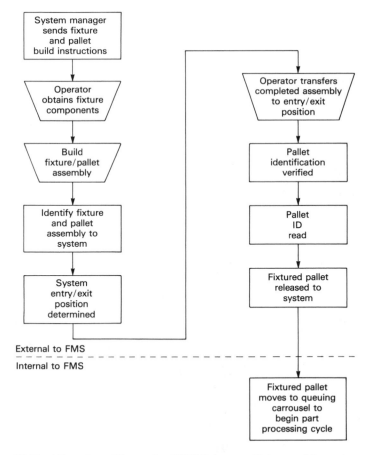

Figure 15-20   Flow chart illustrating FMS fixture–pallet assembly, entry, and exit cycle.

SUMMARY

1. Fixtures take on an increasingly important role in an FMS environment because of the increased machine tool utilization rates and round-the-clock operations.

2. The primary function of any fixture is to locate and secure the workpiece for succeeding machining operations.

3. Part accuracies requiring work-holding fixtures depend on the accuracies of the fixture itself.

4. Fixtures must be designed for ease of part loading and unloading, cutting tool-to-part accessibility, maximum rigidity with supported clamping, and easy access for chip removal.

5. Fixtures should be foolproofed to prevent incorrect part-to-fixture and fixture-to-pallet or table orientation.

6. Fixture design should be simple and standard components should be used whenever possible.

7. Overly elaborate fixture designs only increase costs without adding significantly to accuracy or quality.

8. Accuracy of the part being machined is the single determining factor as to what the tool or fixture tolerance should be (20 to 50 percent of the part tolerance).

9. The tool designer must know the various types of locators and how each should be used to get the best workpiece placement with the least number of locators.

10. Locational inaccuracies develop because of the difference in position and location tolerances between the fixture and workpiece.

11. Modular fixturing employs the use of reusable, interchangeable components that can be easily assembled and torn down based on changing part requirements.

12. Modular fixturing affords users many advantages over dedicated fixturing, ranging from reduced manufacturing, inspection, and rework costs to reusable components that can be easily, repositioned, or changed on a quick turnaround basis.

13. The basic strategy when designing fixtures for FMS application is to complete as many machining operations as possible in one clamping of the workpiece.

14. In an FMS, loading the fixtures is done internal to the FMS, but fixtures must still be designed to minimize part handling time.

15. Specific loading and clamping instructions may be required for loading and clamping fixtured parts in an FMS environment.

16. In an FMS, some fixtures must be married or pinned to a particular pallet to minimize interchangeable error buildup.

REVIEW QUESTIONS

1. Which one of the following is not a requirement of good fixture design?
   (a) Cutting tool and part loading and unloading accessibility
   (b) Reduced fixture-to-table and part-to-fixture setup time
   (c) As many clamps and locators as possible to hold and locate the workpiece
   (d) Utilize design simplicity and use of standard components

2. Workpieces can be distorted under clamping pressure if clamps are unsupported and placed on the fixture at any convenient location. True or False?

3. To _____ a fixture design, parts should be able to be located and clamped only one way.

4. Fixture designers do not need to consider cutting tool length requirements to determine optimum fixture-to-table and part-to-fixture positioning. True or False?

5. All fixtures should be designed and made as simple as the part design permits. True or False?

6. As a general rule, the tool or fixture tolerance should be between _____ and 50 percent of the part tolerance.

7. _____ of the part being machined is the single determining factor as to what the tool–fixture tolerance should be.

8. The types of supports used to locate work from a flat surface which set the vertical position of the part are:
   (a) Solid supports
   (b) Adjustable supports
   (c) Equalizing supports
   (d) Balancing supports
   (e) All of the above
   (f) Only (a), (b), and (c)

9. The terms "locators" and "supports" are never used interchangeably when discussing the devices used under a fixtured workpiece. True or False?

10. Locational inaccuracies can arise from locating a fixtured part from both its previously machined holes and _____.

11. Locating a part from a previously machined hole is not the most effective way to accurately position the workpiece. True or False?

12. The concept of _____ fixturing employs the use of reusable, interchangeable components that can be easily assembled and torn down based on changing part requirements.

13. _____ pins locate the part in a holding fixture, while _____ pins prevent movement but permit ease of part loading and unloading.

14. List four advantages of modular fixturing over dedicated or hard fixturing.

15. The basic strategy when designing fixtures for FMS applications is to complete as many machining operations as possible in one clamping of the workpiece. True or False?

16. Although FMS pallets and fixtures are designed to be interchangeable, some fixtures must be pinned to particular pallets because:

   (a) Inaccurate edge or tongue slot locators are used.

   (b) Cumulative error buildup may exist and interchangeability increases the likelihood of error stacking or accumulation.

   (c) Operators have difficulty keeping fixtures and pallets together and organized in an FMS.

   (d) Pallet and fixture identification is easier.

17. In an FMS, loading and unloading fixtures can be done internal to the FMS cycle (while the complete system is in operation). True or False?

Part V

FMS Computer Hardware, Software, and Communication Networks

The computer first appeared on the factory floor in the late 1950s in the form of the first numerical control units and punched tape. The idea of operating several machine tools from a remote computer was then introduced in the mid 1960s as the first DNC application. NC and DNC where the first of many computer applications to the whole technology of manufacturing. Today computers are used for a variety of applications, many of which are manufacturing related. These include NC programming, inventory control, inspection, process planning, and tool design. And a vast array of software (computer programs and routines), interface networks, and telecommunication links now permeate the entire manufacturing arena.

Three main issues account for the rapid growth and development in computer use:

1. Rapid breakthroughs in computer hardware and microprocessor technology
2. Large-scale price reductions in hardware costs
3. Advanced and increasing capabilities of application software

As the need for interfacing computers and functions together has increased, so also has the complexity due to the accelerated proliferation and diversity of equipment, software, and local area networks (LANs). The FMS computer, software, and communication lines to the various components and workstations are vital to the system in order to gather data and control, activate, and make decisions about rapidly changing system activities.

Computer use in an FMS installation, through its application software, should not be seen as an end in itself, but as an important link to CIM and the concept of closed-loop factory automation and management.

16

System Hardware and General Functionality

OBJECTIVES

After studying this unit, you will be able to:

- [ ] List the general functions performed by computers in a manufacturing environment

- [ ] Name the principal computer hardware elements in an FMS installation

- [ ] Understand the primary functions of programmable controllers and identify where they are used in an FMS

- [ ] Describe the general functionality of a cell controller

- [ ] Identify the different types of communication networks and understand how they are used to transmit data

INTRODUCTION

FMS computer hardware is the visible computing element in a system installation. Overall computer hardware includes the central FMS computer, its related peripheral equipment, programmable controllers, and a backup computer, in some cases, for traffic and/or material management.

The computer can do nothing, however, without the required application software, people, and the necessary communication links to the various workstations. The computer requires proven application software, competent and trained personnel, and backup resources in order for the entire system to perform at acceptable levels.

It is important to remember that the FMS computer and its overall architecture must be given careful consideration and attention during the planning phase, just like the other system components. Without a detailed approach to correctly size the computer in relation to its executable application software, the tasks to be performed, and future expandability, too much or too little computing power may be purchased for the system.

The FMS computer is a tool and functional component like any other element in a flexible manufacturing system. Although it is simply a means by which the FMS application software executes and initiates system activity, it is, in essence, the heart of an FMS.

GENERAL FUNCTIONS AND MANUFACTURING USAGE

The computer continues to gain wide acceptance in all aspects of business and personal applications. Various types, sizes, and styles (Figures 16-1 and 16-2), along with a wide range of price and computing capability, are offered. In many cases, potential purchasers are confused by such an ever increasing array of hardware availability and options.

However, computers and computer purchases should be application dependent. That is, what the computer is being purchased to do, depending on how

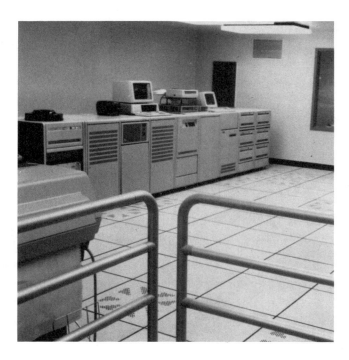

Figure 16-1   Various types, sizes, and styles of computers are available for a wide range of applications. (Courtesy of Cincinnati Milacron)

Figure 16-2 **A wide range of prices, options, and computing power is also available for computers. (Courtesy of International Business Machines Corp.)**

broadly or narrowly defined, should be the driving force behind the actual purchase. In a manufacturing organization, for example, many activities surround computer usage, such as payroll, accounting, engineering, manufacturing engineering, and shop data collection. These activities involve not only the day to day transactions but also management information for analysis and decision making.

Typically, in a manufacturing operation the computer is used to store, process, retrieve, manage, and control data relating to the following functions:

1. Sales records and forecast data
2. Accounting, payroll, and cost control information
3. Design engineering data consisting of drafting, analysis, revision history, and bills of material
4. Shop floor control
5. Tool inventory and design information
6. Work-in-process inventory and scheduling control
7. Capacity planning and process planning
8. NC programming
9. Shipping and receiving data
10. Quality control information

These data groups, although listed as independent pieces of information and designed for convenience of the user, must be able to interface through one form or another and exchange files and records.

In some cases this information is captured on several different computers in many different data bases and in varying formats. In other cases, the various information resides on one central computer and in one data base. This is the centralized–decentralized argument of computers and computer control. Each has its advantages and disadvantages.

The first large mainframe computers to control manufacturing operations were expensive and every attempt was made to fully utilize their centralized computing power. It quickly became apparent that, as installation size and complexity increased, reliability and responsiveness decreased. In the case of a major computer failure, if the entire system breaks down, all plant communications would be lost. User change requests to application programs processing on a mainframe are funneled through a central software group, evaluated with respect to other changes, prioritized, and placed in queue, all of which take time and contribute to user delay and eventually user apathy. However, having large central computers provides more consolidated control of computer charges and expenses, while reducing duplication of new application programming effort within the overall organization.

Decentralized computers (those handling pockets of applications within an organization) give users more control of their own destiny for improved responsiveness and may be connected to other computers or to a central mainframe computer for data distribution. Duplication of programming effort may exist as users provide for their own needs. Additionally, centralized control of expenses and computation charges is difficult to obtain due to local or departmental control of the decentralized computers. The chart in Figure 16-3 displays the principal differences in a centralized versus a decentralized computer approach.

With the rapid increase in microprocessor technology, large-scale price reductions, and increased capabilities of application software, personal computers are more and more making their way to the factory floor (Figure 16-4 on page 354). These personal computers are in many cases networked together for on-line real-time communication and may be tied to a host computer for central data base control of common and shared information. A typical manufacturing architecture and related applications can be seen in Figure 16-5 on page 354. Applications could range from manufacturing monitoring and control to assure that schedules are met and activities are coordinated, to processing of NC part programs and DNC download of tape data files to machine tool controllers. Other applications for this type of networked architecture include shop floor labor reporting, material dispatching, and production, tooling, and inventory control.

In smaller shops, powerful personal computers handle a multitude of these tasks due to the increased capability and affordability of modern-day computers. Computers, monitors, and terminals in most cases are hardened (sealed from harsh shop environment contaminants) to accommodate the rapid increase in shop floor applications.

The computer has gained and continues to gain rapid use in manufacturing as

CENTRALIZED	*DECENTRALIZED*
High hardware cost	Lower hardware cost
High software cost	Lower software cost
High in-plant wiring and connection costs	Lower in-plant wiring and connection costs
Software complex, time consuming, and difficult to change and maintain	Software application specific, designed for local use, easy to write, modify, and maintain
Easy to trace overall operating costs and control expenses	Harder to track overall operating costs and control expenses
Low computer transaction response time, priorities assigned and controlled by corporate data processing	Fast computer response time; user controls own application evironment and assigns own priorities
Expandability difficult and hard to justify cost	Expandability easy, less expensive, and less difficult to justify
All plant communications shut down with computer failure	Only isolated location shut down with computer failure
Expensive backup resources required	Less expensive backup resources required; spare computer can be available as standby
Application program changes time consuming to implement; central data processing must evaluate, prioritize, analyze, and determine system impact	Application changes easy to make and in control of local users

Figure 16-3   Principal differences in a centralized versus a decentralized computer environment.

well as other business and personal environments. The application and required response time will dictate the type, size, architecture, and ultimately the cost of the system, whether centralized, decentralized, or clustered together.

HARDWARE CONFIGURATION AND CONSIDERATIONS

Generally, a hierarchy of computerized functions automatically operates and monitors the entire FMS or any other automated manufacturing system. The devices in

Figure 16-4   Personal computers are more and more making their way to the factory floor. (Courtesy of International Business Machines Corp.)

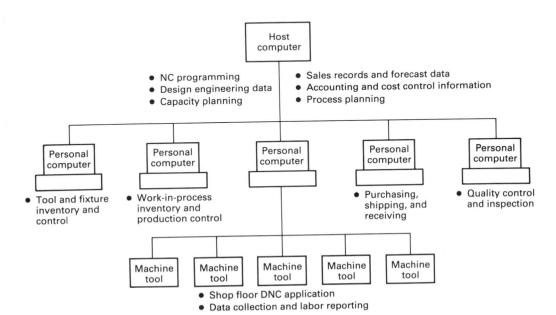

Figure 16-5   Typical manufacturing computer architecture and computer applications.

which these functions reside include the computer numerical control units for each machine tool (Figure 16-6), the host CPU (central processing unit, Figure 16-7), and the minicomputers (programmable controllers, Figure 16-8), which send and receive signals and coordinate system operation. These computers are integrated through a communication network providing bidirectional communication between the appropriate levels of the function hierarchy.

Other hardware elements that comprise a typical FMS or manufacturing system include:

1. Magnetic tape drive (Figure 16-9 on page 357) used to process various magnetic tapes for system backup
2. Expansion cabinet (Figure 16-10 on page 357), which contains the necessary cables and boards that connect the CPU to the peripheral equipment, the other computers, and the machine tool controls
3. Paper tape punch and reader (Figure 16-11 on page 358) used to process part program tape load to the FMS and punch tape when required for input to the MCUs in the event of system shutdown
4. Disk drives and cabinets (Figure 16-12 on page 358) used as permanent storage devices for the operating system, the FMS application software, and all files needed to operate the FMS software

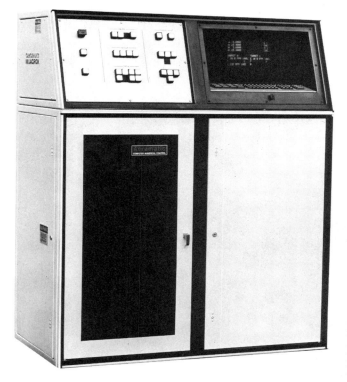

Figure 16-6   A CNC's microprocessor is part of the overall computer hierarchy in a cell or FMS. (Courtesy of Cincinnati Milacron)

Figure 16-7   The host computer is the highest level of control in an FMS. (Courtesy of Remington Arms)

5. System console (Figure 16-13) used by the system manager to activate the FMS software and to display diagnostic messages and messages from system operators

6. Line printer (Figure 16-14) used to produce hard-copy printouts of various management and status reports for the FMS

Figure 16-8   Programmable controllers send and receive signals in an FMS as part of the computer hierarchy. (Courtesy of Allen-Bradley Co.)

Figure 16-9 Magnetic tape drive unit. (Courtesy of International Business Machines Corp.)

Figure 16-10 Expansion computer cabinet that contains the necessary cables and boards that connect the CPU to the peripheral equipment. (Courtesy of Remington Arms)

Figure 16-11 Paper tape punch unit. (Courtesy of Numeridex, Inc.)

7. Video display terminals (Figure 16-15 on page 360) used to issue and communicate FMS commands that run FMS software and display text and graphic output

Excluding the video display terminals that are on the shop floor for operator communications, all the other computer hardware equipment is generally located in an elevated computer room (Figure 16-16 on page 360) overlooking the shop floor. This serves to:

1. Provide the system manager with a view of the entire FMS from the system console

Figure 16-12 Disk drive unit. (Courtesy of International Business Machines Corp.)

Figure 16-13   FMS system console–control room. (Courtesy of Remington Arms)

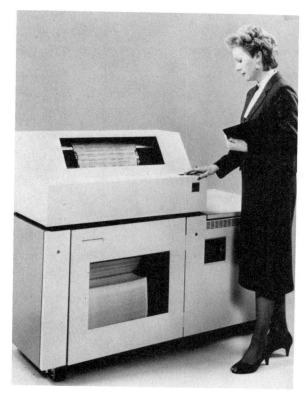

Figure 16-14   Line printer unit. (Courtesy of International Business Machines Corp.)

Figure 16-15   Typical video display terminals (VDTs) for communication. (Courtesy of Remington Arms)

Figure 16-16   Elevated computer room overlooking the FMS equipment. (Courtesy of Cincinnati Milacron)

2. Protect the computer equipment from potential shop floor hazards, such as flooding and, fork trucks or careless use of overhead cranes
3. Provide a centrally located, environmentally controlled room for all sensitive computer equipment immediately accessible by the system manager
4. Save valuable floor space

Additional features of an overall FMS computer architecture provide:

1. Industrial local area networks (data transfer and communication lines)
2. Distributed, fault-tolerant architecture
3. Expandability to accommodate additional processing requirements and data
4. Gateways to other plant computers
5. Data-base management system for controlling data and information stored in the system data base
6. Various programming language capabilities

Difficult cellular or FMS components and applications require the availability and careful selection of the entire computer architecture. For this reason, all computer elements, including the communication network, must be given careful consideration and tailored toward the specific application. Factors that help determine the overall architecture include:

1. Size and complexity of the installation
2. Degree of automation
3. Physical layout of the plant
4. Real-time data requirements and response time
5. Number of hierarchical control levels

The FMS computer control system manages all system components that contribute to the automatic production process. Generally, FMS computer systems include:

1. Sufficient nonvolatile storage to simultaneously store part programs, executive programs, utility programs, and maintenance programs
2. Capability to multiplex (transmit information from several datalines into a single line) part programs to all CNC controllers
3. Standard operating system
4. Application software that has been developed and tested on the specified computer hardware
5. Standard, real-time communication interfaces and terminal access to the host mainframe (not the FMS host computer) from all FMS workstations
6. Storage, retrieval, and archiving of all status information; this information is either monitored in real time or retrieved at a later time for display on the FMS control room console

7. Direct, dedicated communication lines between the FMS central computer and the machine tool controllers capable of error-free part program transmissions

Considerable planning and forethought must be given to computer hardware purchases. In some cases the computer hardware equipment may be purchased by the overall FMS supplier as part of the turnkey FMS system purchase, or the host computer, PLCs, tape drives, and other peripheral equipment may be purchased independently by the system user, linked together, and integrated with the other system elements. Determining how much computing power is actually required is based on system size, tasks to be performed, response time required, and future expandability, which are the critical elements. Potential expandability should be both vertical for adding larger and increased processing capabilities and horizontal for clustering or adding another processor for distributed growth.

The following user considerations should be kept in mind with respect to computer hardware purchases for an FMS:

1. Decide between a turnkey purchase through the system supplier as part of the overall system or "do it yourself." "Do it yourself" applications may create more headaches, utilize more experienced human resources than anticipated, and not save any money in the long run. Turnkey purchases hold the system supplier responsible for system hardware maintenance and performance through the acceptance testing and warranty periods.

2. After purchasing the specific computer hardware, the dependency on that computer vendor will continue for the life of the application.

3. The amount of equipment from different computer hardware vendors should be limited. Multiple vendor hardware linkage and compatability may be difficult, require unanticipated interface software, and delay startup. In many cases, FMS application software is written and will process on only one vendor's computer.

4. The FMS computer will not run unattended. Consequently, knowledgeable and resourceful system managers must be selected to staff and manage the computer resources and the entire FMS. This usually means one system manager for each operating shift.

5. The system hardware should not stand alone. If not initially, the FMS computer should eventually be tied to other computer systems supporting MRP (material requirements planning), the inventory data base, CAD (computer-aided design) systems, and other factory communication systems. Consider short- and long-term integration compatibilities.

6. Computers, PLCs, and peripheral equipment will ultimately require service. During the warranty period and acceptance testing in a turnkey purchase, this is the system supplier's responsibility. Maintenance contracts can be established with the hardware vendors to supply service as needed after the initial warranty period. Strive to become more self-reliant through personnel development and training of backup resources.

7. The life of a computer system depends on the application and growth of the user's needs. Therefore, the life expectancy of FMS hardware is based on whether new and/or extra applications and requirements can be accommodated by the present system and its support devices without sacrificing response time or other possible trade-offs.

PROGRAMMABLE LOGIC CONTROLLERS

Programmable controllers, often referred to as PCs or PLCs (programmable logic controllers), are electronic devices or small computers that are used to control machinery, actuate devices, and assist in the control of advanced automation systems. PLCs are the technological replacement for electrical relay systems that are rapidly replacing the hard-wired circuits that have controlled the process machines and driven equipment of industry in the past.

The function of a PLC, as seen in Figures 16-17 and 16-18, is to examine the status of an input or set of inputs and, based on this status, actuate or regulate an output device or devices. Input to a PLC may be either discrete or continuous. Discrete PLC inputs typically come from photocells, proximity and limit switches, push buttons, microswitches, and pressure switches. Continuous PLC inputs come from voltmeters, potentiometers, or thermocouples. PLC outputs are directed to actuating hardware, such as solenoids, solenoid valves, and motor starters, and in the case of FMS to initiating some activity at the various workstations.

Figure 16-17   Typical programmable logic controller (PLC). (Courtesy of Allen-Bradley Co.)

Figure 16-18   Programmable logic controller (PLC) knowledge and familiarity are important to cell or system operating and maintenance personnel. (Courtesy of Allen-Bradley Co.)

A PLC is composed of four primary elements.

1. Central processing unit (CPU)
2. Power supply
3. Memory
4. Input and output (I/O) signal-handling equipment

Generally, the CPU is a microprocessor and supplies the brainpower for the PLC. Most PLCs now being offered are microprocessor based and have more logic and control capabilities than the earlier electronic logic circuit models. The CPU scans the status of the various input devices continuously, applies the input signals to the memory control logic, and produces the required output responses needed to activate and control the equipment or workstation entry–exit points.

PLCs are generally grouped by size based on the number of I/O handling capabilities, functional attributes, and memory capacity. Micro and mini PLCs are usually modern replacements for relay systems. Larger units may have the functional capabilities of a small computer and be able to handle computational functions, generate and output reports, and provide high-level communications capabilities.

Instructions are input to a PLC in the form of programs, just as for other computers. Four major programming languages are generally used with PLCs. These include ladder diagrams, Boolean mnemonics, functional blocks, and English statements. Some PLC systems even support high-level programming languages such as BASIC and Pascal.

The PLC's function in an FMS is to receive signals to monitor and activate the I/O queues of each work–load station, as seen in Figure 16-19. This includes each of the individual workstations for part processing, inspection, cleaning, and others. Additionally, PLCs are used to control load–unload and fixture build stations, queuing stations and carrousels, automatic storage and retrieval systems (ASRS), and control coolant–chip reclamation systems. Signals are passed back and forth between each of the PLCs in the FMS and the host computer in order to activate and verify pallet shipment, movement, and registration and receipts and to initiate activity of other FMS system functions.

CELL CONTROLLERS

Cell controllers (Figure 16-20) are devices responsible for the coordination of multiple workstations, machines, or operations that offer information and communications processing and coordination capability. They combine the capabilities of PLCs and minicomputers.

Figure 16-19   PLCs receive signals to control, monitor, and activate input–output queues for the various workstations. (Courtesy of Remington Arms)

Figure 16-20   Typical cell controller. (Courtesy of Allen-Bradley Co.)

Cell controllers are generally factory hardened to exist on the shop floor but are not used to directly control shop equipment. Cell controllers generally are used to control PLCs or PCs, which in turn control a manufacturing cell or a series of machine tools, as seen in Figure 16-21. Cell controllers provide computerized supervision and coordination of multiple controllers along with data collection and concentration for the factory floor. Cell controllers also provide a distributed data base and communication capabilities to higher-level computers, such as the factory host or the inventory control system computer.

Cell controllers are generally used for smaller-scale FMS or cellular systems where the full range of system decision-making capabilities is not required to support diverse part mix and lot size requirements. This would include mid- to higher-volume applications with some part type and mix variety but high cell coordination and data management requirements.

The primary difference between cell controllers and PLCs is the computer language and knowledge required to program and maintain them. Little computer knowledge is necessary to program or operate PLCs. Cell controllers, on the other hand, require some degree of computer knowledge, along with more operator knowledge and training to use than programmable controllers.

Repair parts for PLCs, if easily replaced, may be stocked in some cases by electrical repair personnel or they may be obtained from distributors. Cell controller maintenance or repair generally requires skilled in-house engineering skills

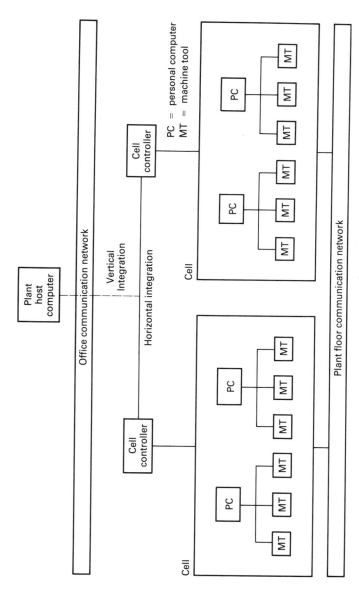

Figure 16-21 Cell controller architecture illustrating vertical and horizontal integration.

or the original equipment manufacturer's (OEM) field service technicians. How quickly a manufacturer can react to cell or system trouble is an important consideration when selecting a cell controller vendor.

Communications between various cells, other plant computers, and the factory floor can be either horizontally or vertically integrated (Figure 16-21 on the previous page). Communication between the islands of automation or manufacturing cells is horizontal integration and should be considered as a primary building block of CIM. This level of integration may be sufficient if automation is the only goal. This level of communication establishes the foundation for vertical communications.

Vertical communications are integrated upward and downward between the plant host computer and office level and the cell and plant floor level. However, it is important to note that, without horizontal communication integration, vertical integration cannot be fully achieved in a CIM network.

One big factor relative to cell controllers is what data need to be managed. Project team members must identify the necessary data and functions that the cell controller will handle. The project manager should make sure all levels of management and the project team make a thorough analysis of what the cell controller's input will be and exactly what data need to be managed and controlled.

Many cell controller applications do not require the functionality or the price of larger systems. Smaller-scale cell controllers, for example, can be used to automate toward CIM in a logical step by step or phased in approach. Such an approach can hold down the price of distributed cell controller architecture, thereby lowering the overall implementation cost of CIM.

COMMUNICATION NETWORKS

Communication networks are the information highways of an automated manufacturing system. Selection of either the network or the computer, in many cases, may determine the other. Some networks are closer to being standardized and supported by computer vendors than others.

Networks are generally localized based on the elements that need to be linked together in a given area. Consequently, the acronym LAN (local area network) is used in many cases to designate the network or data transfer line. Local area networks (LANs) fit between the computer connection and long-haul networks such as the telephone system. Local area networks may be limited to a room, a building, an automated system, or a series of closely connected systems or buildings.

Network topology is the road map of the entire network. Although the word topology is basically a misuse of the word topography, it is the geometric layout of the data links and computers that require linkage. Network topology can have many forms, but the two most common are point to point and multidrop.

Point-to-point topology is a circuit connecting two points or computer nodes without passing through an intermediate point. As seen in Figure 16-22, its primary use is for very simple or subnetworks. A multidrop network (Figure 16-23) is a

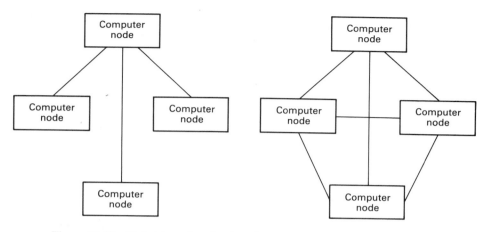

Figure 16-22   Point-to-point circuit connecting two points without passing through an intermediate point.

single line that is shared by two or more computer nodes. Multidrop networks reduce overall line costs but increase the complexity of data transfer in the network, as well as the cost of line connection. The method of data control and priorities in either a point-to-point or multidrop application is the control topology.

Networks are generally classified in three types:

1. Star or radial
2. Ring or loop
3. Bus

Control of a star or radial network (Figure 16-24) remains at the node where two points are joined. The main controlling node may also be called the master or net master node. The connecting point would be called the slave. This is a simple master–slave relationship.

Ring or loop networks may be classified in two types depending on control

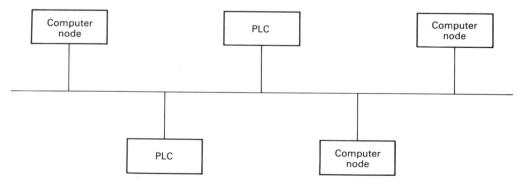

Figure 16-23   A multidrop network shares a line with two or more points.

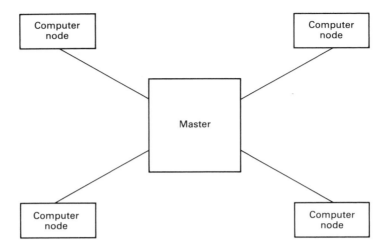

Figure 16-24   Star or radial network controlled by the net master in a master–slave relationship.

type, as seen in Figure 16-25 on page 371. Centralizing control in one node of the network creates what is generally referred to as a loop network. Subnodes in the network can only communicate with other subnodes when permitted by the controlling master node.

Ring networks use distributed control. In this case, each node can communicate with every other node without direction from a controlling master node. This method is more complex than a loop arrangement, but in either case (ring or loop) data may be passed from node to node around the ring. Each node must have an active repeater to transmit the data to the next node.

One of the most commonly used forms of control topology is the bus network, as seen in Figure 16-26 on page 372. A bus network is significantly different from the other arrangements in that data may be sent to all nodes at the same time, as opposed to passing data from node to node around a ring.

The efficiency of a bus network in an FMS or any other automation application depends on the following factors:

1. Reliability, fault tolerance, and availability
2. Data transmission rate and maximum distance between nodes
3. Time delay to respond to interrupts and data requests
4. Geographic distribution of components requiring node connections

Channel access is a means of determining who controls the network. Polling is a technique in which each node's access to the network is determined by the master node.

If a centralized polling scheme is used, the central node will query each subnode and ask if it has access to the network. Each node's frequency of access depends on how much other data traffic needs to be passed between the other

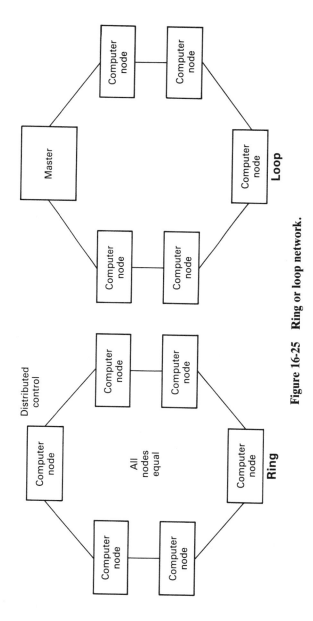

Figure 16-25 Ring or loop network.

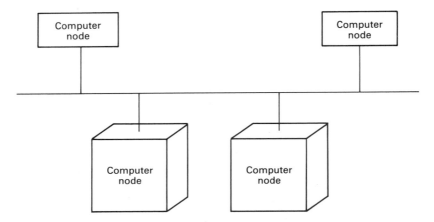

Figure 16-26   In a bus network, data can be sent to all nodes at the same time.

nodes in the network and the total number of nodes on the network. Some polling schemes can assign priority to selected nodes by querying them more often.

A common form of distributed polling is called token passing. Token passing (Figure 16-27) is generally associated with ring or loop networks and functions by passing a packet of bits called a token around the loop until it reaches a node that requires access to the network. That particular node will grab and hold the token while it sends its data. Once a message is on the ring, it is passed from node to node

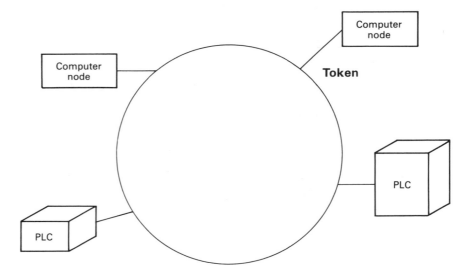

Figure 16-27   Token passing is a common form of distributed polling.

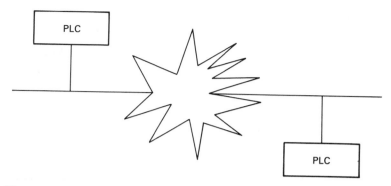

Figure 16-28   Collision detection prevents two nodes from attempting to take control of the bus at the same time.

until received by the destination node. Messages usually circulate back to the sending node to confirm receipt. When the sending node has completed its transmission, it puts the token back into circulation.

Another form of polling is collision detection, as seen in Figure 16-28. It is used primarily with bus networks where information is sent to all nodes at the same time and all nodes have equal access to the bus network. Collision detection prevents two nodes from attempting to take control of the bus at the same time. A computer node, ready to transmit, senses if the bus is busy and can sense any change in the energy level of the bus. It interprets this as a warning of a possible collision with other existing network data. Upon detecting a possible collision, each node involved backs off and waits a brief fixed or random period of time and then retries the data transmission.

Signaling techniques used in connecting cable for communication networks can be broadband or baseband. Broadband cable has multiple channel access like cable TV. Multiple signal transmissions occur at the same time, but only a few channels are used at a time in a particular home. Broadband networks usually run slower than baseband networks, but support a variety of data channels.

Baseband cable has only one channel where the only signal on a baseband cable is the network traffic. The primary advantage of baseband over broadband is the interface simplicity. It has only one channel to decode, and problems with cross channel interference generally do not exist.

Figure 16-29 illustrates the communication network and connections in an FMS. Although considerable emphasis has been placed on network standardization in recent years, such as General Motors manufacturing automation protocol (MAP), wide acceptance of various imposed standards has been slow to be adopted.

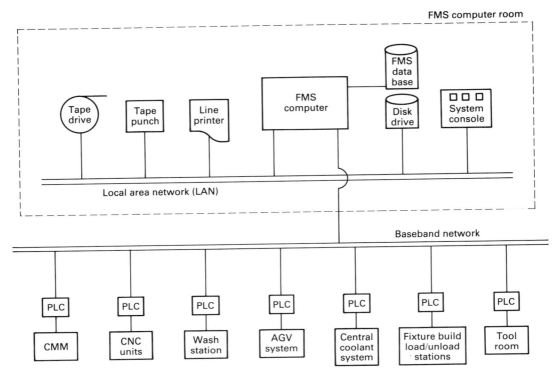

Figure 16-29   Communication network connections in a typical FMS.

SUMMARY

   1. FMS computer hardware is the visible computing element in a system installation and is the heart of an FMS.

   2. Some computer applications in a manufacturing environment include design engineering drawing information and bills of material, capacity and process planning, shop floor data collection, NC and DNC, inventory and scheduling information, and shipping and receiving data.

   3. There are distinct advantages and disadvantages for centralized and decentralized computer approaches.

   4. In some smaller shops, powerful personal computers handle a multitude of manufacturing tasks due to their increased capability and affordability.

   5. A computerized hierarchy automatically operates and monitors the entire FMS.

   6. Computer hardware elements and FMS equipment are integrated through a communication network providing bidirectional communication between the appropriate levels of the function hierarchy.

7. The FMS host computer, magnetic tape drive unit, expansion cabinet, tape punch, disk drives, system console, line printer, and video terminals all make up the FMS computer hardware elements.

8. The life of a computer system depends on the application and growth of the user's needs.

9. Programmable logic controllers are electronic devices (small computers) that replace electrical relay systems and are used to control machinery, actuate devices, and assist in the control of advanced automation systems.

10. The PLC's function in an FMS is to receive signals to control, monitor, and activate the input–output queues of each work or load–unload station.

11. Cell controllers are devices that are responsible for the coordination of multiple work stations, cells, or machines.

12. Communication networks are the information highways of an automated manufacturing system.

13. Networks are generally localized based on the elements that need to be linked together in a given area.

14. Network topology is the road map of the entire communication network.

15. Networks are generally classified in three types: (1) star or radial, (2) ring or loop, and (3) bus.

16. Polling is a technique in which each node's access to the network is determined by the master node.

17. A common form of distributed polling is called token passing.

18. Signaling techniques used in connecting cable for communication networks can be broadband or baseband.

REVIEW QUESTIONS

1. Having more control over computer charges and expenses is a strong case for a decentralized computer approach. True or False?
2. The devices in which the computer hierarchy functions reside include the host CPU, the minicomputers (PLCs), and the _____ .
3. Factors that help to determine the overall FMS computer architecture include:
   (a) Size and complexity of the installation
   (b) Degree of automation and physical plant layout
   (c) Real-time data requirements and response time
   (d) All of the above
   (e) Only (a) and (b)
4. Multiple vendor hardware purchases should be limited because linkage and compatibility between the various units may be difficult and require unanticipated interface software. True or False?
5. Input to a PLC may be either continuous or _____ .

6. A PLC is composed of four primary elements. These are:
   (a) CPU, power supply, software, and I/O equipment
   (b) Memory, CPU, software and communication network
   (c) Power supply, memory, I/O equipment, and CPU
   (d) CPU, memory, I/O equipment, and bus connectors

7. The PLC's function in an FMS is to receive signals to control, monitor, and activate the input–output queues of each work and load–unload station. True or False?

8. A _____ network is a single line that is shared by two or more computer nodes.

9. List three types of classifications generally used for networks.

10. Collision detection is a form of polling. True or False?

11. Signaling techniques used in connecting cable for communication networks can be _____ or _____ .

12. Cell controllers are not factory hardened as they are primarily located in an office environment. True or False?

17

FMS Software: Structure, Functions, and Description

OBJECTIVES

After studying this unit, you will be able to:

- ☐ Discuss the importance and applicability of software to an FMS installation
- ☐ Describe how typical FMS software is developed and structured
- ☐ Identify principal FMS software modules and briefly explain their function
- ☐ Understand what simulation is and how it is used in an FMS installation
- ☐ List the principal FMS user and supplier software warranty and security issues

INTRODUCTION

An FMS, as we have already seen, is made up of many different elements. (The major elements consist of a variety of processing, quality assurance, computer hardware, and system support equipment, all of which are visible and tangible.)

(Software is an invisible element and is the essential glue that binds the visible FMS equipment together and forms a system.) Without these highly developed and sophisticated computer routines, an FMS is a mere collection of individually automated equipment on the factory floor. It is the FMS software that actually drives the entire system, calling the various equipment to action through command-driven operator and system manager input. If the FMS computer is the heart of a flexible manufacturing system, the FMS software is its life blood.

In this chapter, general and generic FMS architecture and descriptions will be presented in a simplified manner in order for students to understand the overall structure, function, and capability of an FMS's most vital element.

GENERAL STRUCTURE AND REQUIREMENTS

Software is the vital invisible element that actually drives the FMS. There are two basic levels of software required for an FMS: (1) operating system and (2) application software. Operating system software is the highest level, is computer manufacturer specific, and executes supervisory control over the application software. Application software is usually developed and supplied by the system supplier and includes all the FMS specific programs and routines.

Application software for an FMS is complex, highly proprietary, and for many companies, represents several hundred worker-years of development effort. Generally, it is composed of several modules, each of which is made up of a series of computer programs and routines relating to various functions performed within the system. These include NC part program download from the FMS host computer to machine tool controllers, traffic and material-handling management, work-order generation, workpiece scheduling, simulation, and tool management. All these software modules must be well designed and function predictably, reliably, and interactively in order for the FMS to perform at peak operating efficiencies and acceptable levels. Poorly designed software prevents manufacturers from achieving the full flexibility and potential capacity of FMS.

FMS software, because it is the life blood of a flexible manufacturing system, is also the most complex, least understood, and strategically important aspect of an FMS. Structured and coded properly, tested rigorously, and functioning adequately, it can make an FMS productive at unprecedented performance levels. It should be added that all completed FMS software can only be considered acceptable after it has been thoroughly checked out with the system in complete operation in the customer's plant.

Although all the individually developed FMS software modules must ultimately interact together to run the entire system, there are three main advantages of modular software design and development:

1. Segmenting the required software into modules for development permits simultaneous efforts once overall system requirements and specifications are agreed on, thereby saving time
2. Modular software design and development permits phased installation, allowing the user to begin using the system while some portions are still in the development, testing, or implementation mode

3. Quicker and easier tracing and pinpointing for problem, enhancement, or modification changes

The model depicted in Figure 17-1 is a generic example of FMS software structure and is by no means indicative of FMS software architecture for all suppliers of these systems. The modular structure, however, does provide a simplified approach to learning fundamental FMS software capabilities. Figure 17-1 illustrates the FMS computer and software modules and their appropriate connections to processing equipment and operator stations on the factory floor. A comparison to gain perspective can be made to a direct or distributed numerical control system (DNC) (Figure 17-2), where the DNC computer controls NC program download to CNC controllers, remote job entry (RJE) upload for part program processing, tool management, and management reporting information. The DNC portion is only a subset of a complete FMS software system. Individual software modules will be described in more detail in the next section.

The real-time, multitasking FMS software allows for many simultaneous activities to take place. For example, the system manager could be running the planning aids or starting one or more machines while production control personnel could be inquiring into work-order status and NC data are in the process of being transferred to the FMS computer from magnetic tape.

In general, FMS software permits the following activities and functions to be performed within the system:

1. Access to critical data for user-customizable management and status reporting
2. Operational data collection
3. Scheduling and simulation of system activities
4. Workpiece load balancing for efficient utilization of processing equipment and tooling
5. Remote distribution of NC programs to machine control units
6. NC part processing at a remote site via telecommunication links
7. NC program library control and maintenance
8. Automated tool and workpiece delivery
9. Management of fixture, pallet, and tooling data
10. Definition and assignment of workpiece operational sequences (routing)
11. Automated workpiece inspection
12. System control of workpiece entry and exit
13. Overall control of system resource utilization
14. Error diagnostic output and display for system maintenance and troubleshooting

For any of these functions to take place within an FMS, certain requirements must be met for the various components in an FMS to function together as a system. These requirements include:

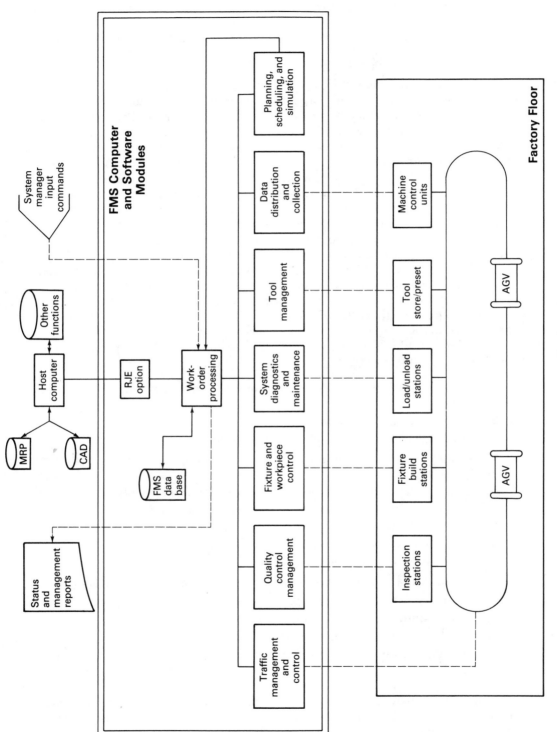

Figure 17-1 General FMS software structure.

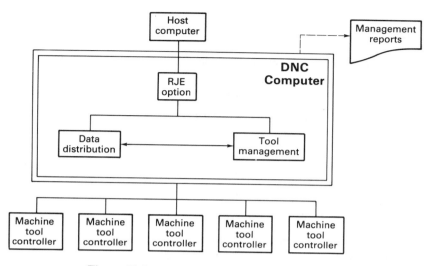

Figure 17-2   Typical DNC system architecture.

1. Complete installation and functionality of all FMS components and data lines
2. Thoroughly tested and complete operational software
3. Completed and proven NC and CMM part programs
4. Assembled and proven fixtures
5. Available tooling and material
6. Fixture and pallet assignment and identification
7. Completed work-order creation

Modularity of software design does not necessarily imply that all systems using the same or similar software modules are created equal. Many FMS users have highly specific and esoteric requirements to suit their own applications and operating concerns. Some of these might include specific FMS software modules to couple an already existing automatic storage and retrieval system (ASRS) to a new FMS or to have the FMS directly receive production requirements and part scheduling information from the host computer.

Overall, FMS software, like other types of computer software, is as different and autonomous as the people who develop and code it. What counts is what it does and how well it performs in a manufacturing environment.

FUNCTIONAL DESCRIPTIONS

(FMS software performs a variety of functions and activities that are normally invoked by commands entered by authorized system users at various terminals connected to the system.) Extensive monitoring, reporting, and updating of a common data base are done on a dynamic and real-time basis by various sensing devices and operator input throughout the system.)

(The FMS software, although written as a series of computer programs and in modular form, functions collectively and uniformly in full system operation.) To better understand the broad functional capabilities of FMS software, each module presented in Figure 17-1 will be examined in more general detail.

1. Work-Order Processing

The primary function of this software module is to be the vehicle by which various workpiece production schedules and processing sequences are implemented, controlled, and managed within the system. It provides the software capability to create and structure the routing of workpieces to be run through the system and to manage, control, and dispatch workpieces and operational resources within the FMS. It is the principal software element through which the others are dispatched and utilized.

Work orders must be created (generally by system manager input commands at a terminal in the computer room, as seen in Figure 17-3), just as conventional process plans or routings are created for processing a specific workpiece. All processing steps must be listed in sequential order, and the resources to be utilized, such as part programs, cutting tools, and fixtures, must be available and accessible. Work-order preparation must be completed well in advance of the part's scheduled run time. Any workpiece introduced into an FMS must be identified to the system by means of a work order. Identification for each workpiece type generally includes defining the number of parts to be processed, start date, due date, and

Figure 17-3   Work orders and other system commands are input to the FMS computer by the system manager. (Courtesy of Kearney and Trecker Corp.)

routing sequence. Work orders must then be authorized by the system manager for processing activity to begin.

The objectives of a work-order processing and workpiece control software module are to:

- ☐ Define an individual work order to the FMS and describe its station processing sequence

- ☐ Initiate entry and exit of workpieces within the system

- ☐ Reserve system resources to ensure production and completion of in-process work orders

- ☐ Start and stop processing of work orders in response to changing production requirements, bottlenecked machine groups, or machine breakdown situations

- ☐ Establish and control multiple workpiece processing sequences and priorities for parts competing for the same resources at the same time

- ☐ Permit alternate routing creation and system entering for parts due to changing production requirements or catastrophic failures

- ☐ Collect and report current and historic information regarding completed and in-process work

- ☐ Interact with the various other FMS software modules to provide for workpiece movement and delivery and other resource availability and allocation (for example, tools, fixtures, and part programs)

2. Data Distribution and Collection

The software written for this module basically sends data to (downloads) and receives data from (uploads) the machine control units. The primary function is to retrieve NC data from FMS computer disk storage and transfer it to the requesting machine control unit (MCU) as required. Other functional capabilities include:

- ☐ Storage and retrieval of specific or cumulative cutting tool, machine tool, or other data collection information

- ☐ Security provisions to avoid sending a part program to an incorrect machine tool; to allow only authorized access to NC data files, and protection for the integrity of data in the system; additionally, authorized users can gain access to only "proven-out" and current part programs for downloading to MCUs

- ☐ Data manipulation, prioritization, and file management capabilities for NC data transfer

- ☐ Real-time status of system resources and management report information; reports of this type are meant to show the current status of the system with respect to production, tooling, and other system resources for short-term management decision making

- ☐ Send messages from a terminal or MCU in the system to any or all terminals in the system

- ☐ Uploading of APT source files from a remote site via the RJE option and transmission of the postprocessed NC output back to the FMS computer for storage and later use

□ Logging of all system transactions along with time of day and user identification

3. System Diagnostics and Maintenance

This module provides diagnostics for scheduled and unscheduled maintenance and troubleshooting of system components. In some cases a systematic menu-driven sequence of diagnostic and recovery instructions is displayed to the system manager in the event of a catastrophic failure or shutdown. These instruction sequences are also accessible to system operators and inform them of diagnostic steps and checkpoints in order to help pinpoint problem areas.

The diagnostics and maintenance module will:

□ Collect data stored at the MCUs to maintain a historical maintenance and cumulative run time log
□ Inform the system manager of regularly scheduled preventive maintenance for all system components
□ Notify the system manager when tool failures occur
□ Detect and identify system failures, and notify the system manager when a failure is detected
□ Verify specific failures through comparison using a library of diagnostic programs
□ Communicate via a telephone line with a vendor's computer for remote diagnostics capability
□ Monitor the FMS equipment using feedback sensors in the material-handling system and processing stations
□ Maintain up to date program libraries and diagnostic routines for maintenance and troubleshooting of system components and facilities

4. Tool Management

The tool management software module is responsible for storing, managing, and updating cutting tool data files. It provides the capability to assign the necessary tooling to the various processing stations in the system. Additionally, it allows for entering, in advance, the associated data for the required tooling, along with the ability to transport and load that tooling at the designated machine tools.

Tool management software generally will:

□ Store, manage, and update tool data files containing tooling data, which typically consists of the tool length, cutter diameter compensation, feed-rate and spindle overide information, and tool cycle time expectancy
□ Add new tool assemblies and tool groups to the system
□ Delete completed tool groups or time-expired tool assemblies from the system
□ Authorize tool movement to and from the workstations
□ Display, via a system terminal, the complete list of tools for a specific part program

- Display specific tool or tool group status information
- Search the tool data file for specific tool groups or tool assemblies
- Automatically enter tool data from tool-gaging equipment
- Permit editing of tooling data by authorized users
- Address and identify the specific machine tool and tool pocket that holds each tool
- Identify the pockets in each machine's tool matrix from which tools have been exchanged (for loading and unloading purposes)

5. Traffic Management and Control

The traffic management and control function controls and monitors the movement of tools and palletized workpieces between processing, parking, and load–unload stations in an FMS. It accepts signals from other software modules (principally tool management, data distribution, and work-order processing/workpiece control) and initiates timely action of the material-handling system. These various stations consist of machine tools, queuing carrousels, load–unload stations, wash stations, battery recharge stations, and gaging and inspection stations.

The primary functions of the traffic management and control module are to:

- Issue commands to move tools and palletized workpieces between the various stations in the FMS
- Control an AGV's registration, positioning, and lift mechanism via sensors for pallet pickup and delivery
- Track and maintain current and historical part, pallet, and tool movement data between processing parking stations and load stations
- Provide overall supervisory control (including collision control) of all AGVs in the material-handling system
- Provide operating personnel with software input commands to control AGV traffic

6. Quality-Control Management

Quality-control management software provides the capability to collect, store, retrieve, and archive workpiece inspection data. Current machine tool-cutting data are compared with workpiece specifications. Any deviation outside the part tolerance band will direct a message to the system manager and cause resultant action to begin.

The quality-control management module will:

- Control and direct the coordinate measuring machine (CMM) and compare inspection results with previously input workpiece tolerance specifications; workpiece inspection programs are created at the CMM and uploaded to the FMS computer for storage and retrieval
- Check that machining stations are processing parts to required specifications
- Identify the specific part, call for a download of the appropriate inspection program to the CMM, and initiate the measurement cycle

□ Store and archive inspection data, prepare the measurement report, and notify the FMS computer of the completion of the measurement cycle

□ Insure that only the most recent inspection program is stored on the FMS computer to avoid the risk of downloading a previous version

□ Associate each measurement program with the correct workpiece and specific routing suboperation for matching when retrieval is initiated

7. Fixture and Workpiece Control

The fixture and workpiece module works in conjunction with the work-order processing module. It is used to control the status of all fixtures and parts known to the system. Workpiece loading and unloading of fixtures are performed at one or more fixture stations by system operators. Operators use terminals (Figure 17-4) to communicate to the FMS computer relative to workpiece and fixture processing readiness. A CRT message will instruct the operator to either store or disassemble a fixture and pallet assembly.

Other functions performed by the fixture and workpiece control module would be:

□ Authorize fixture load–unload completion for workpiece entry and exit

□ Validate workpiece-to-fixture and fixture-to-pallet identity

Figure 17-4   Operators use terminals to communicate with the FMS computer relative to workpiece and fixture control. (Courtesy of General Dynamics Corp., Fort Worth Division).

□ Control and monitor fixture and workpiece activity from system entry to exit

□ Initiate the material-handling system for workpiece movement to and from processing stations

□ Permit operator communication with the FMS computer and system manager as to load and unload completion and workpiece processing availability

□ Assign a fixture identification to every fixture and fixtured pallet used in the system

□ Dispatch workpieces that have failed CMM inspection to a material review station for reinspection and disposition

8. Planning, Scheduling, and Simulation

This module provides the user with the capability to do production planning and workpiece scheduling in advance of actual due dates and to simulate (mimic) the results. It provides for the selection and designation of system resources, and to schedule work and see the system impact under "what-if" conditions without initiating actual workstation activity. This is an extremely powerful and useful software module that can be of immense benefit to system users as they attempt to plan, schedule, and balance load and production requirements.

This module will:

□ Provide predictable results of work-order processing based on prior definition of production requirements and needed system resources

□ Provide the capability to assign operation sequences and tooling for processing operations on the various workstations in the FMS

□ Simulate results of production load situations based on input of work-order release times, pallet allocations, and other pertinent input data

□ Verify that existing or proposed production schedules will achieve the required throughput and results

□ Display system element utilization levels and provide production capacity data

□ Determine the most efficient workpiece batching and scheduling strategies

□ Ensure the availability of system resources to achieve expected production rates

□ Obtain useful simulated results quickly by modeling the system under load conditions and analyzing its performance

□ Analyze the impact of system expansion (increasing the number of workstations or AGVs, for example) and increasing and/or varying production requirements

This is a powerful and useful software module, particularly with respect to simulation. Because of its capabilities, usefulness, and increasing acceptance and use among manufacturers, simulation will be discussed separately later in this chapter.

OPERATIONAL OVERVIEW

The overall functional capabilities of an FMS are much too broad and complex to be presented here in great detail. However, now that individual software modules have been discussed, a generic operational overview can be presented and better understood. An operational flow chart, depicted in Figure 17-5 on pages 390–391, parallels the following discussion to enhance understanding of the functions controlled by FMS software.

Before any FMS can be operated at full capacity, all its component parts must be installed, operational, checked out, and able to function together as a complete system. NC and CMM part programs, fixture, tooling, and other required resources, must also be tested, approved, and available.

The system manager, who is a first-line supervisor having responsibility for operation of the system, enters control commands at an FMS computer terminal to start the machines, devices, and remote terminals for the day's activities. System shutdown would also be performed from the system manager's terminal. System security measures generally password-protect unauthorized personnel from entering FMS commands and initiating action.

Workpiece production is controlled by the creation and management of work orders and routing sequences. Work orders generally must be created and present in the system prior to any workpiece entry. Workpiece entry into an FMS generally requires the following steps to take place:

1. System resource start and checkout
2. Verify workpiece and measuring part program identity
3. Group similar processing stations
4. Add and identify required pallets and fixtures
5. Assign and match fixtures to pallets
6. Assign required tooling
7. Create detail part work orders (routings)
8. Approve and authorize the work order
9. Start work-order processing (system entry)
10. Stop work-order processing (system exit)

Once work orders are created for the required parts, they must be authorized (usually by the system manager). Authorization of a work order is required in order to:

1. Ensure that the right parts and the right quantities are manufactured consistent with the master production schedule and due date requirements
2. Guarantee availability of system resources to produce the workpieces associated with the work order

After authorization, the first machining destination is selected and the work order is started. If the destination station is blocked (perhaps another part is there still being machined), the pallet may be moved to the queuing carrousel for overflow parking.

Tool assemblies are created for each workpiece required within a fixed production window based on tool lists created when the part programs were written. Tool groups are scanned (bar code read for inputting tool dimensional data) and dispatched to the appropriate machine tool for loading in each machine's tool matrix. Tools are then checked and validated that all are correct, present, and loaded and that tool dimensional data matches the tool list file in the FMS computer for that specific part program, workpiece, and operation.

Part fixtured pallets are delivered to the load queue of the appropriate machine tool and the fixture-to-pallet identity is verified (to prevent a wreck by machining the wrong part with the right tools and NC program).

The NC part program is downloaded to the machine control unit where the part is to be processed, and it is verified to be correct and current. The fixtured pallet is then shuttled into the machining zone and the machining cycle begins. The MCU controls the entire machining process until completion of the program unless some form of failure or operator intervention interrupts the cycle. If, for example, the machine senses a tool failure or another machine system failure, the machining cycle stops and the system operator and system manager are notified via a machine fault signal. The operator must take corrective action at the machine or if the fault cannot be corrected, abort the operation, and take the machine out of production.

When the machining cycle is complete, the pallet is shuttled to the unload queue where it is then picked up by a dispatched AGV and transported to the load queue at the wash station. Pallet identity is read and checked for tracking purposes, and the wash station selects the predetermined wash cycle type and the cleaning cycle begins.

Following completion of the wash cycle, the pallet is shuttled to the unload queue and awaits departure to the inspection station. The pallet is transported to the load queue at the inspection station and its identity is verified. It is then shuttled into the inspection position for measurement, and the CMM inspection program is downloaded and checked for its correct identity and latest revision level.

The CMM program begins and inspects the part or parts on the fixtured pallet. Measurements are compared with acceptable part tolerances, the results are archived, and an inspection report is generated. If completed inspection results indicate out of tolerance deviations, the pallet is transported to a material review stand for inspection. The machine that produced the rejected part then produces another for detailed inspection. If similar results are obtained, the machine producing unacceptable parts is taken out of production and the problem is immediately investigated. If the part (or parts) passes inspection, the pallet is shuttled to the unload queue and subsequently transported via AGV to a load–unload station. The part or parts are then unloaded or refixtured for the next series of processing operations.

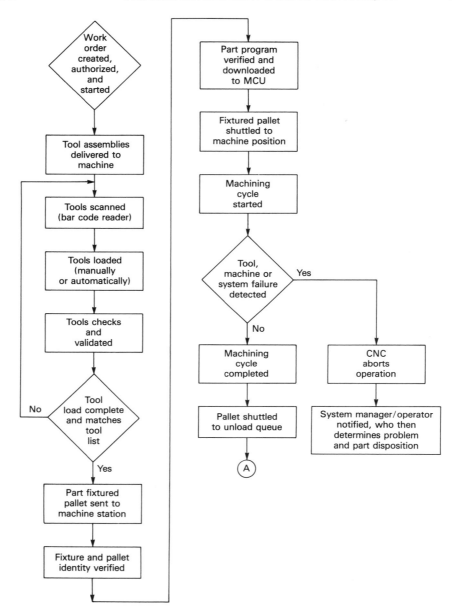

Figure 17-5   Operational overview depicting part flow through an FMS.

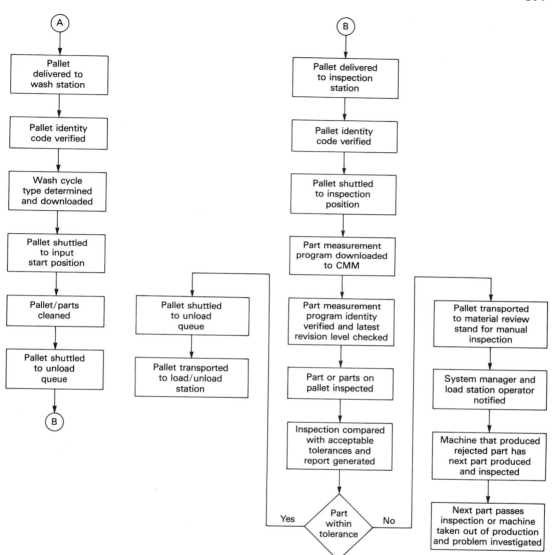

Figure 17-5 (*Continued*)

COMPUTER SIMULATION

Simulation is a powerful analytical software tool that is used to mimic the behavior of a system. With simulation, a mathematical model is built of a particular process that allows users to observe the outcome of manufacturing operations without experimenting with the actual equipment. It is a simplified image of the actual process. Simulation, particularly relative to FMS, affords users the opportunity to know the consequence of system changes before they happen through the powerful routine's predictive, look-ahead, and what-if capabilities. Computer simulation has become widely recognized and accepted as a fast and accurate analytical tool for studying complex manufacturing systems and operations. Developing alternatives and examining system and production impact by manual methods would normally require hundreds of calculations and extensive human resources.

Simulation languages may be classified according to different concepts, activities, or capabilities, which are:

1. Event oriented
2. Process oriented
3. Transaction or translation oriented
4. Activity or function oriented

Simulation results may be viewed through both textual and/or graphical means on a computer terminal.

Simulation, because it is a software tool consisting of known and available resources, activities, and alternative methods, permits close analysis of the manufacturing process to be examined. Data developed in one phase of the simulation process are passed as results to the next phase of the simulation analysis in order to find the optimum result. Figure 17-6 illustrates the general phases of a simulation analysis.

The most important advantage of simulation is the almost unlimited predictive variations of the actual process that can be generated and examined for possible selection. And simulation analysis can be done independent of other ongoing computer functions and without affecting existing operations. It should be cautioned, however, that simulation techniques as powerful and capable as they are, should not be a sole substitute for planning. Simulation should be used as an aid to augment, test, and verify planning activities.

Simulation analysis, in many cases, is performed by the system manager on the FMS computer to develop, model, and examine alternatives in order to:

1. Optimize production schedules
2. Maximize resource utilization
3. Meet multiple part production due dates
4. Maintain production requirements in the event of individual workstation downtime problems
5. Increase production

6. Balance machine load
7. Balance tool requirements

Additionally, simulation may be used as an aid to plan a new facility or to change an existing one, to search for a more efficient machine layout, to observe traffic and material congestion, and to test the effect of workstation failures. Through computer simulation, the consequence of these and other changes and alterations can be reviewed and the best alternatives selected to meet system and production goals and objectives.

Problems can develop in the use of simulation based on assumptions and interpretations. For example, a user can graphically view pallet travel paths on a CRT screen and the interaction of various system workstations. The user can

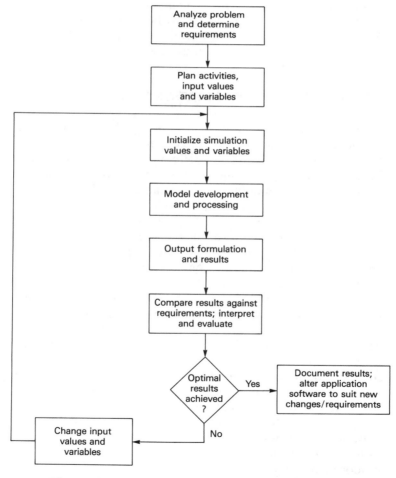

Figure 17-6   General phases of a simulation analysis.

assume that pallets are moving and work queues are building at the travel speed viewed on the screen. In actuality, however, the pallet traverse rate may be slower or faster than that actually displayed through the graphic simulation, and an analytical verification should be made. Also, users may assume a simulation report's output accuracy by the number of digits after the decimal point. Users must be aware that simulation report output data accuracy is only as good as the input data accuracy. An accurate simulation will represent the type and accuracy of its input data.

One of the most critical aspects of simulation has to do with capability testing during the development versus acceptance phase. Obviously, development testing must occur, but assuming success of functionality at this stage is unsupported and inconclusive. Erroneous assumptions and incomplete data may have been used. Consequently, bad (and costly) decisions may be made as a result of reliance on simulation software that has not been thoroughly tested under actual operating conditions in the user's plant. Simulation software and all FMS software should be rigorously checked and evaluated by both user and supplier systems' personnel. And it should be final tested with the system in operation as part of the overall system acceptance procedure in the user's plant.

WARRANTY, SECURITY, AND INTEGRATION ISSUES

Operating system software, as part of an FMS or any computer installation, is developed and maintained by the computer manufacturer. Warranty periods for mainframe computers are generally one year from date of computer installation. Extended warranty or maintenance contracts are written at the time of purchase or following expiration of the initial warranty period.

FMS software is generally developed by the system supplier or the system supplier contracts the programming work to be done to a software vendor. In either case the system supplier owns the FMS software. In some cases the FMS user may contract to purchase the FMS software. Specifically, this means the source programs (source programs are computer programs written in a symbolic language such as FORTRAN or COBOL) are purchased. In this case the user would own the source programs and have access to the various routines once the entire system was formally accepted. Object programs (the internal machine language equivalent of a source program that is converted from a source program by means of a translator routine) are what actually execute the software in a computer. Without the source programs, however, it is almost impossible to make alterations to the software. In instances where the supplier wants to maintain its proprietary interests, object programs are supplied with the system to execute the FMS software while the supplier holds and maintains source programs and documentation.

Because of the proprietary and extensive capability of FMS software, warranty and security issues are a high priority for both system user and supplier. System users want to avoid software-related problems, such as the software vendor going out of business and not being able to alter or service the code, and preventing unauthorized employees from using the system. System suppliers want to protect their extensive software development effort from being copied and from being altered by eager but inexperienced users. The chart in Figure 17-7 depicts both user and supplier possible software warranty and security problems, along with preventive measures or remedies. FMS software is extremely complex and interrelated. Arbitrary and guess-work software changes with the FMS in full operation can cause problems that seriously affect system performance and impair error tracing and detection. Consequently, both system users and suppliers are very protective and security conscious regarding software authorization, accessibility, and changes.

USER		SUPPLIER	
Problem	**Prevention/ Remedy**	**Problem**	**Prevention/ Remedy**
Unauthorized employee system access	Password protect system entry, assign user identification, and log transactions	Potential for copying proprietary software	Do not supply source programs; supply only object programs
Software problems during warranty period	Supplier debugs and tests	User buys source-makes coding changes that cause difficult to trace ripple effects in other routines	Do not sell source; keep in escrow account and update as required
Software problems after warranty period	1. User learns and maintains system or 2. Contract supplier to fix		
What if supplier goes out of business (supplier owns source)?	Keep copy of source in escrow account (negotiate during contract discussions)		
User wants to make own software changes and/or additions	Warranty voided; user assumes all software maintenance responsibility		

Figure 17-7   FMS software general warranty and security issues

Some companies purchase an FMS as a stand-alone cell or system with limited or no plans to expand or to integrate the FMS computer and software with other business systems. Other companies are concerned with how the FMS software can be integrated with other existing or added-later systems. This would imply a connection between the FMS computer and the host computer, as seen in Figure 17-1 on page 379, or perhaps other computers operating as a functioning part of the business.

Many potential users of FMS will spell out the integration of FMS software with existing other systems as part of the requirements and specification document. This lets the FMS supplier and software vendor know early in the planning and requirements phase exactly what other computer systems the FMS computer and software must be linked to.

Generally, the primary reasons to integrate an FMS computer system to a central host computer are to:

1. Download material requirements planning (MRP) information directly to the FMS computer and avoid manual determination and entry

2. Process NC part programs from an FMS terminal on the host computer and direct output results via the remote job entry (RJE) feature

3. Interface with other management information systems (MIS) for data exchange, decision making, and report generation

4. Interface with existing or expanding CAD/CAM systems to provide more globalized computer-integrated manufacturing (CIM) capabilities

Many purchasers either know specifically at the time the FMS purchase is planned what existing computer systems the FMS computer system should integrate with, or they want the integration provisions embedded in the FMS software to allow for expandability and integration at a later date. Perhaps users may want to add an automatic storage and retrieval unit (ASRS), another cell, or a different type of FMS later. Many users of automated cells and systems see the capability of these vital linkages to pass data back and forth between computer systems as a key business objective and the opportunity for sweeping productivity improvements. It is important to note that many companies do not see automated cells and systems as an end in themselves, but as a means to achieving an end—a means toward achieving CIM.

SUMMARY

1. FMS software is the life blood of an FMS and is what actually drives the entire system by calling the various equipment to action. It is the most complex, least understood, and strategically important aspect of an FMS.

2. Two basic levels of software are required for an FMS:
   (a) Operating system software, which executes supervisory control over the application software

(b) Application software, which includes all the FMS-specific software programs and routines

3. Poorly designed software prevents manufacturers from achieving the full flexibility and potential capacity of FMS.

4. FMS software is highly proprietary, requires several hundred worker-years of development effort, and is usually developed and written in application-type modules.

5. FMS software performs a variety of functions and activities that are normally invoked by commands entered by authorized system users at various terminals connected to the system.

6. Before any FMS can be operated at full capacity, all its component parts must be installed, operational, checked out, and able to function together as a complete system.

7. Simulation is a powerful analytical software tool that is used to mimic the behavior of a system. It affords users the opportunity to know the consequence of changes before they happen.

8. The most important advantage of simulation is the almost unlimited predictive variations of the actual process that can be generated and examined for possible selection.

9. FMS software is generally developed by the system supplier or the system supplier contracts out the programming work by be done to a software vendor.

10. Software warranty and security issues are a high priority for both system user and supplier.

11. Many FMS purchasers want to integrate the FMS computer with their existing host computer or other systems to be added at a later date for data exchange.

REVIEW QUESTIONS

1. The two basic levels of software required for an FMS are operating system and _____ software.

2. Modular software design and development enable the user to begin using the system while other modules are still in the development, testing, or implementation mode. True or False?

3. FMS software permits:
   (a) Operational data collection and control of system resources
   (b) Management of fixture, pallet, and tooling data
   (c) Remote distribution of NC programs to machine control units (MCUs)
   (d) Only (b) and (c)
   (e) All of the above

4. The FMS software module that is normally responsible for NC program download from the FMS computer data base to the MCUs is the _____ module.

5. The tool management software module is responsible for:
   (a) Storing, managing, and updating tool data files containing tooling data
   (b) Searching the tool data file for specific tool groups or tool assemblies
   (c) Issuing commands to move tools and workpieces between the various stations in an FMS
   (d) Only (a) and (b)
   (e) All of the above

6. The FMS software module that is normally responsible for creating and routing workpieces through the system is _____ .

7. Generally, pallets and fixtures in an FMS do not need to be matched, assigned, or identified. True or False?

8. If the part or parts on a particular pallet pass inspection at the CMM, the next step is to be transported to the wash station. True or False?

9. Explain the basic functional capabilities of computer simulation.

10. It is necessary to final test all FMS software as part of the overall system acceptance procedure in the user's plant. True or False?

11. A primary user warranty and security issue is:
    (a) Copying of source programs
    (b) Who actually makes software changes and/or corrections
    (c) Unauthorized employee system access
    (d) None of these

12. Integrating an FMS system and software with a host computer for direct download of material requirements planning (MRP) information should be given consideration by the user during the planning stage. True or False?

Part VI

FMS Installation and Implementation

During the FMS installation and implementation phase, a multitude of activities are happening at the same time. Internal and external plant construction is well underway, processing equipment is being installed, and in-plant and vendor personnel are working side by side. It is also a time when schedules are being challenged and contingency plans are paying off.

This is a critical time for the project. FMS purchaser, supplier, and vendor personnel are all onsite and people are working longer hours to maintain what is normally an aggressive installation and implementation schedule. The purchaser–supplier courtship, marriage, and honeymoon phases are over as relationships become strained and tempers sometimes flair during this high-pressure time. It is a time of technical, operational, and interpersonal challenge.

The installation and implementation period is also a time of great exhilaration and feeling of accomplishment, as all the discussions, preparations, meetings, and planning sessions are now yielding visible and tangible results. Both customer and supplier stand at the threshold of completion and success. Things are really starting to come together and happen. However, the installation and implementation period can also be a sobering experience as it is a true test of how well all the previous planning and preparation work was done.

FMS Installation

OBJECTIVES

After studying this unit, you will be able to:

- ☐ Identify and describe the general installation sequence for an FMS
- ☐ List some potential problems related to system installation and start-up
- ☐ Describe what training should take place before, during, and after FMS installation
- ☐ Discuss some key factors that can help maintain a healthy purchaser–supplier relationship

INTRODUCTION

During the FMS installation phase, the system must physically be assembled, started, and debugged. Also during this period, some training will occur while the system is being installed, and some usually occurs after the system is installed and in complete operation.

Direction and communication are extremely critical elements during this period due to the high activity level. People must know and understand the installation plan and schedule, and those who will work as part of the FMS team should begin to work together. Working together at this early stage alongside vendor personnel enables the FMS team to learn as much as they can about the equipment and will help to cultivate system ownership and teamwork.

Problems and delays are sure to be encountered during this final but critical stage of the project. However, these can be resolved with well-prepared contingency plans, a teamwork ''can-do'' attitude, and project leadership as well as project ownership.

INSTALLING THE SYSTEM

Although the type, size, and complexity of an FMS will determine installation requirements, most system installations follow a predetermined and structured sequence of events. There are no hard and fast rules for FMS installation. However, logical and orderly occurrences must take place, which are jointly determined and agreed on ahead of time by the system purchaser and supplier.

Many purchasers prefer a phased installation approach that allows them time to bring various pieces of equipment on-line and in some cases into production as quickly as possible, as opposed to waiting for the entire system to be installed and become operational. Customer requirements, scheduling demands, and numerous other timing considerations must be kept in mind during installation.

As with any automation, construction, or other major project, certain elements may be completed at the same time, while others must be completed following a rigid sequence of events. With FMS, installation can be modular yet parallel. A typical FMS installation that might be accomplished in four phases would be:

Phase 1 (Figure 18-1): Site excavation and central coolant and chip removal and recovery system installation would occur during this early stage. Figures 18-2

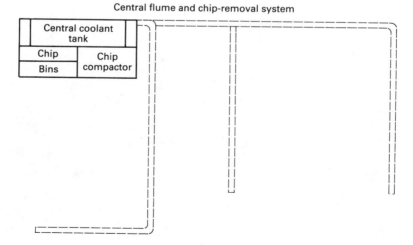

Figure 18-1   Phase 1 installation completion.

Figure 18-2 **Installing the chip, flume, and coolant system. (Courtesy of Remington Arms)**

Figure 18-3 **Site excavation and construction. (Courtesy of Remington Arms)**

Figure 18-4 Completed and cured concrete FMS manufacturing floor. (Courtesy of Remington Arms)

and 18-3 illustrate these early and important preparatory steps. Underground utilities would be laid, and chip bins and a chip compactor installed if required. Machine foundation slabs and the main FMS manufacturing floor would be reinforced, and concrete poured and cured (Figure 18-4).

Phase 2 (Figure 18-5): At this time, AGV lines are cut into the floor from the floor cut layout and floor marking templates. Guidepath wire is embedded in the floor cuts and epoxy sealed. The elevated computer room is erected, additional utilities and conduit are run to appropriate locations, and wiring is pulled. Compressed air and electrics are run to machine power drop locations (Figure 18-6), and the overhead crane system is installed. Additionally, the central coolant flume and chip recovery system, chip conveyor and chip compactor are run off and operationally checked.

Phase 3 (Figure 18-7): During this phase, machines that have been transported by truck to their destinations are unloaded and placed in their proper positions. Electrical and compressed air lines previously run to drop locations are hooked up. Machine tools are quickly aligned, and operational features (machine and control) are thoroughly checked out (Figure 18-8 on page 406). Machine cutting tests, generally conducted, witnessed, and approved by the purchaser project manager in the machine tool vendor's plant, are conducted again once machines are installed in the customer's plant. Again, cutting tests are witnessed and approved by the purchaser project manager. The FMS host and AGV computer system is also installed during this phase, and the operating and application soft-

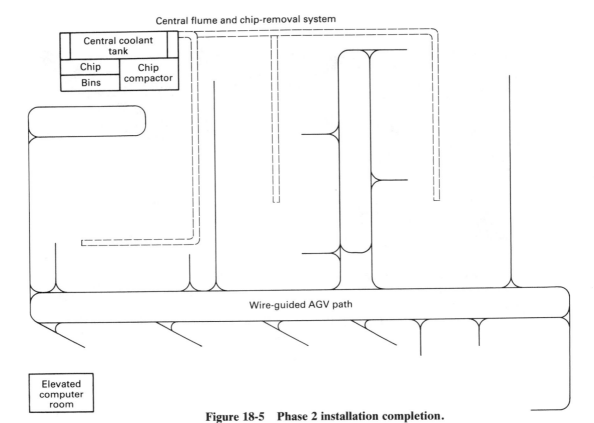

Figure 18-5   Phase 2 installation completion.

Figure 18-6   Compressed air and electrics are run to machine power drop locations. (Courtesy of Remington Arms)

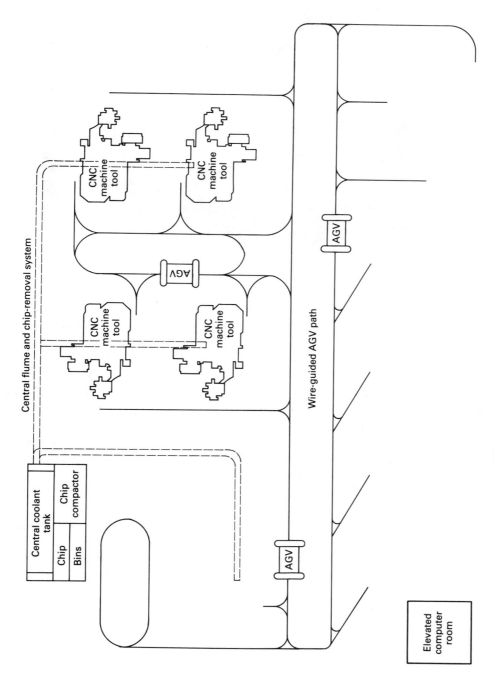

Figure 18-7   Phase 3 installation completion.

**Figure 18-8   Installed machines and controls are thoroughly checked out.
(Courtesy of Remington Arms)**

ware is installed. Installing the FMS computer and software at this stage permits early computer system testing and debugging. AGVs are installed, checked out, operated manually, and linked to the computer system.

Phase 4 (Figure 18-9): Adding the inspection machine is one of the next most important modules to add to the FMS. Other modules quickly added to the system, many of which are occurring at the same time, include installation of the queuing carrousels for overflow parking (Figure 18-10 on page 408), load–unload stations, fixture and tool build stations, and the tool delivery module. If an automatic storage and retrieval system (ASRS) is included, this could be installed sometime during this phase. Also during this phase, the purchaser would be installing perimeter guarding, eye wash stations, fire extinguishing equipment, and other facility-related equipment and safety items, and painting guide path markings on the factory floor.

As each FMS machine or module is added to the system, its performance and conformance to specifications are checked out and verified on an individual or modular basis before linking it with other modules or elements of the FMS. Such rigid modular verification can help eliminate specific problem determination confusion once the entire system is being tested for functionality and performance capabilities.

SYSTEM START-UP AND PROBLEM CONSIDERATIONS

No magic formula exists for an FMS system start-up once all the individual modules and components have been installed. As each FMS machine or module is

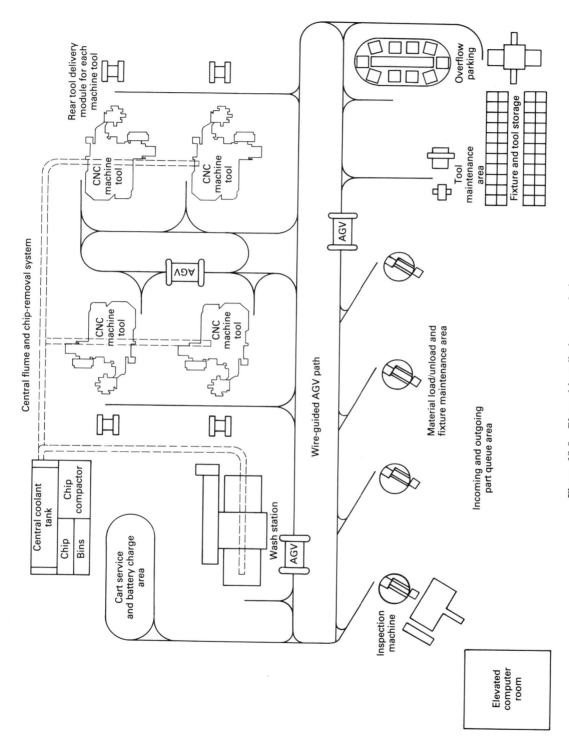

Figure 18-9 Phase 4 installation completion.

407

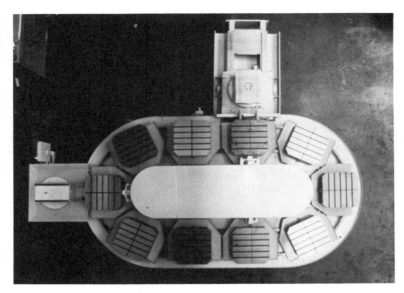

Figure 18-10   Installed and operational queuing carrousel. (Courtesy of Remington Arms)

added to the system, its functionality and performance must be thoroughly checked out before linking it to the system. Such a modular "shakedown" of each FMS component cannot be overemphasized.

Once the FMS application software is installed on the FMS host computer, the various machines can then be checked for functionality and responsiveness as related to the FMS software modules (Figure 18-11). Such stringent check-out of each software module relative to its specific task or function is the FMS supplier's responsibility. And now that each FMS machine or component has been installed in the customer's plant, performance deviations become readily apparent. The system supplier continues checking out, debugging, and fixing all FMS hardware and software elements (Figure 18-12) until the system performs like a "system" and conforms to the functional specifications.

Timing is also a critical factor with respect to system start-up. Both FMS purchaser and supplier have worked long and hard during the planning phase to establish and agree on a master plan, which established the prioritized schedule, sequence of events, and related timetable. Now that most of the items on the master plan have been completed and the system is installed, only system start-up, debugging, training completion, and customer acceptance remain. However, these final steps can take much longer than anticipated for many reasons, many of which are nontechnical.

Some of the potential problems related to system start-up include:

1.   Communication Breakdowns

During the critical installation and start-up phases, people are doing more and talking less. However, these are precisely the times when communication ties need

Figure 18-11   FMS application software must be thoroughly checked out with respect to responsiveness and system functionality. (Courtesy of Cincinnati Milacron)

Figure 18-12   The system supplier is responsible for diagnosing and fixing FMS hardware and software elements. (Courtesy of Cincinnati Milacron)

to be strengthened through regular (sometimes daily) meetings between system purchaser and supplier teams. Such frequent regularity of status meetings helps to reinforce individual and joint project team goals and objectives and allows for discussion and quick correction of strategic plans if required. And frequent, regular meetings promote teamwork and the "project-must-win" attitude between and among the purchaser and supplier project teams.

2. Employee Turnover

Depending on the size, complexity, and installation and start-up considerations, the time from project planning, justification, and start-up to installation and system acceptance testing can be one to three years. During this time, employees leave their assignments for any number of reasons. These are natural occurrences in the business world and must be managed. However, either FMS purchaser or supplier project team employee turnover can seriously affect project success and timely start-up and completion. Steps can be taken to minimize both purchaser and supplier attrition during the duration of the project through:

- □ Careful screening and selection of key project personnel prior to project start-up; individuals must understand the project goals and their roles
- □ Continuous and genuine management (purchaser and supplier) involvement, as opposed to mere lip service
- □ Providing project personnel with adequate resources and continued encouragement
- □ A well-orchestrated reward, recognition and teamwork program along with special promotions, such as hats, T-shirts, and dinners

3. Anticipated Failures

Many equipment, software, and small technical component failures occur during system start-up. This is where prior training pays off as purchaser personnel should be assisting supplier personnel in system start-up, debugging, and trouble-shooting. Even though it is the supplier's responsibility to make the entire system operational, purchaser personnel should be assisting and using this time as an opportunity to learn as much as possible about system start-up failures.

4. How Will This Affect Me?

Now that the system is installed and operational, the reality of a full-scale operating cell or system begins to set in with the people. Although construction and installation are highly visible, nothing brings the impact of FMS to reality like system start-up. In many cases, FMS affects people and internal support organizations on a much wider scale than originally thought. Education and information about FMS should be continuous and ongoing.

5. Management Changes

Sometime during the duration of the project, purchase upper management changes can occur. Management changes, particularly during installation and start-up, can and probably will bring about changes in expectations about what FMS is and what it can do for the business. Project personnel should take the initiative to educate and inform new management about FMS, rather than waiting and trying to change their minds after the new management has already formed an opinion.

6. Additional Perishable Tool Requirements

Now that the system is beginning to operate and function, perishable tool requirements may need to be reevaluated. Although this should have been done in advance, it is difficult in many cases to make a convincing case for the high machine tool utilization rates expected. Consequently, tool inventory levels may not have increased to support the new FMS production rates anticipated.

7. Material Floor Space Allocation

Because of the production capabilities of manufacturing cells and systems, floor space requirements for incoming, waiting, and outgoing work may be over- or underestimated. System start-up and beginning operation will quickly reveal the accuracy of the estimated material floor space requirements.

8. Pre- and Postoperations

System start-up will also reveal the accuracy of estimated FMS pre- and postoperation requirements. In some cases, external FMS scheduling and/or production control adjustments may need to be made to accommodate FMS production rates. Work flowing to or from the FMS may be too fast or too slow based on previous planning estimates, thereby requiring adjustments or alterations to preceding or succeeding FMS operations.

TRAINING EXECUTION

Training is the single most important personnel investment in FMS. FMS success depends on how well the training has been executed and how well the people have learned. Training success depends on how much emphasis was placed on and how well the master training plan was prepared. Completion of a usable and effective training plan is the most important element to the execution of successful training.

Recall from Chapter 5 that training should take place before, during, and after system installation. Considerable training should already have taken place by this time, much of it in the supplier's and vendor's plants. This training consists of various equipment operation, programming, and maintenance classes. This preliminary training familiarizes various selected FMS team members with basic equipment function, along with operational and maintenance considerations and capabilities.

While the equipment is being installed, the following additional hands-on training is taking place:

1. Mechanical and electrical machine tool maintenance training (Figure 18-13): This occurs through purchaser personnel assisting supplier and equipment vendor personnel in erecting, installing, and starting all new machine tools.
2. Inspection machine and material-handling equipment (Figure 18-14): Detailed knowledge is required of this critical equipment. Purchaser personnel

Figure 18-13   On-site mechanical and electrical maintenance training. (Courtesy of Cincinnati Milacron)

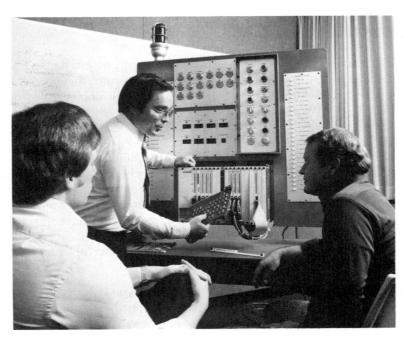

Figure 18-14   Board-level electronic maintenance knowledge is required to service high-tech equipment controls. (Courtesy of Cincinnati Milacron)

must be available to assist supplier and vendor personnel during this installation as well.

3. Queuing carrousel, wash station, and additional support equipment (Figure 18-15): This equipment, although not as technically complex as NC machine tools, AGVs, or CMMs, must be understood mechanically, electrically, and operationally by purchaser personnel. Lack of training emphasis on one element can result in the inability to pinpoint the cause of a problem, thereby shutting down the entire system. Purchaser personnel should be knowledgeable and competent about all system equipment, and the best time to build that knowledge and competence is during system installation and start-up.

After installation, the following training should be conducted.

1. FMS Operation Seminar

Such a comprehensive course or seminar is generally conducted by the system supplier after the system is installed and in complete operation. It is conducted for all purchaser personnel who will work in any capacity as part of the FMS team. This seminar covers the operational aspects of each system element or component individually and the cell or system itself as a complete entity.

Figure 18-15 FMS components must be understood mechanically, electronically, and operationally by purchaser personnel. (Courtesy of Interlake Material Handling Division)

2. FMS Software Seminar

This training requires the complete system in full operation. It is generally offered to the system managers who will be inputting the various FMS software commands at the system console that activates the system. These individuals will have already completed prerequisite FMS host computer classes and are now ready to begin learning and exercising the extensive FMS application software.

3. Mechanical and Electronic Maintenance Seminar

This training also requires that the complete system be in operation. It is conducted by supplier personnel who intentionally create common mechanical and electronic maintenance problems with system components. Purchaser maintenance personnel who have already had several classes in equipment maintenance and assisted with the installation now have the opportunity to test their problem diagnostic and resolution abilities under key supplier personnel supervision. This lab-oriented course is critical to building FMS team competence and confidence in being able to diagnose and solve system problems.

The most successful cells and systems are those where both customer and supplier have worked together as partners to develop and execute the training plan. Additional considerations with respect to training plan execution are:

1. In most cases, training focuses on the technical "what" issues, and sometimes the "why" issues are either completely omitted or left for the FMS team members to assume. Team members must understand the importance and critical nature of their training and their responsibilities to overall FMS success. Individuals conducting the training must raise the level of consciousness and awareness of FMS team members in order for them to focus on the "big picture" of system quality, performance, and productivity.

2. In many cases, it may be necessary to train people on the second or third shift. This could happen due to unanticipated hardware or software technical problems or failures. Or training could possibly be planned for the second or third shift to avoid too many people requiring complete use of the system at one time and to allow more hands-on training time for FMS team members. Such cooperation and flexibility may be necessary during this critical time to take full advantage of a 24-hour day and keep the project on schedule.

3. Purchaser FMS team members should be brought in to assist in installation and other duties as early as possible and kept busy. Team leaders should make sure that training and assisting the supplier and vendor personnel are actually occurring. However, it is the purchasing project manager's responsibility to make sure that all FMS team members across all shifts are adequately trained and functioning as a team.

4. Training does not end with purchaser acceptance of the cell or system. Employee turnover and change are a way of life. System knowledge must be freely passed and exchanged in order to help protect and ensure the success of a sizable investment. Cross-training and backup training should be continuous and ongoing. The FMS project manager and system managers should develop cross-functional training plans and see to it that the continuous training is carried out.

PURCHASER–SUPPLIER RELATIONSHIP

How well the supplier relationship works from system installation through acceptance testing depends directly on the relationship that was developed during the early planning stages and how that relationship matured. It is a marriage for the duration of the project and, like any marriage, requires give and take and a cooperative attitude.

During the critical but final installation through acceptance stages, user and supplier relationships can become strained due to the tremendous pressures of installation, start-up, and acceptance. It is the supplier's responsibility (project manager) to control the vendor's equipment delivery, functional and operating or technical problems, and related training.

The user and supplier project managers are the critical focal points for the relationship. They must maintain a high degree of objectivity during these high-pressure periods and not just defend or protect their own turf. They must demand functional manager ownership, responsibility, and a sense of purpose for each manager's area of expertise. Additionally, project managers must be able to see both sides of a problem or disputed issue and resolve the problem in a rational and timely manner.

Project managers must follow the detailed master plan and be able to guide, steer, and make minor plan adjustments to keep the project on schedule and the user–supplier relationship healthy. Additionally, communication lines and networks must be clearly established and followed. This means keeping all communications flowing through the project managers. This helps minimize confusion, misunderstanding, mistakes, and delays between the user and supplier project teams.

What can be done to strengthen the user–supplier relationship during this time of high pressure and turmoil? Although general in nature, the following guidelines can help keep the relationship and the project on a steady course.

1. Daily meetings should be occurring between the on-site supplier project team and the purchaser project manager and selected team members. Monthly review meetings should occur (preferably at the user's plant) between both purchaser and supplier project managers and selected functional managers, depending on problem or review considerations. Such regular meetings serve to quickly identify and correct problems and foster and strengthen a team relationship.

2. On-site supplier personnel (not necessarily the supplier project manager) must be available on a daily basis (this means live in the purchaser's town) through project completion. Generally, there are field service personnel who handle equipment installation, hook up, and runoff and software personnel who are actively engaged in testing and debugging the FMS application software.

3. A log book should be maintained by the FMS purchaser project manager. This log book would list all problems as they occur (hardware and software), beginning during the installation phase through final user acceptance. Each problem is dated for log-in purposes, an individual is assigned responsibility to resolve the problem by the project manager, and the date is logged in when the problem is corrected. Accurate record keeping via a log book is the most effective way to document and keep track of resolved and unresolved problems, assign responsibility for fixing the problem, and keep track of the time it takes to fix problems. It is also an effective tool for keeping the purchaser–supplier project team relationship on an objective and rational course.

4. In a turnkey FMS installation, many smaller technical problems are turned over to the system supplier for resolution, who either handles the problem directly or has the OEM vendor resolve the problem. This takes the purchaser out of the loop and puts the responsibility on the supplier, where it belongs. However, the user loses a valuable opportunity to learn by turning over the problems to the turnkey supplier and may end up hurting himself in the long run by not staying closer to the problems and maintaining a closer relationship with the turnkey supplier.

SUMMARY

1. An FMS installation sequence must be logical and orderly and agreed on ahead of time by the system purchaser and supplier.

2. Many purchasers prefer a phased installation approach, which allows them time to bring various pieces of equipment on-line and into production as quickly as possible.

3. As each FMS machine or module is added to the system, its functionality, performance, and conformance to specifications are checked out in an individual or modular basis before linking it to other FMS elements in the system.

4. Many of the steps in the final stages of FMS installation, start-up, and acceptance testing take longer than anticipated due to nontechnical reasons.

5. FMS success depends on how well the training has been executed and how well the people have learned.

6. It is important for FMS training to occur at the proper time, some of it before, some during, and some after system installation.

7. The FMS purchaser–supplier relationship is a marriage for the duration of the project, and the success of this relationship depends on how well it developed and matured during the early planning stages.

8. The user and supplier project managers are the critical focal points for maintaining a healthy, objective, and rational relationship.

9. Some of the most effective means for keeping the project on schedule and maintaining a good user–supplier relationship are having daily and monthly status review meetings, keeping a log book of resolved and unresolved problems, making sure communication lines are established and open, and demanding functional manager ownership of project responsibilities.

REVIEW QUESTIONS

1. FMS installation can be modular, yet parallel. True or False?
2. Cutting AGV lines into the floor would be:
   (a) One of the last items completed
   (b) The FMS purchaser's responsibility
   (c) Done after the main FMS manufacturing floor is completed
   (d) Completed when the central flume system is installed
3. Machine tool cutting tests must be approved by the purchaser _____ .
4. The system _____ is responsible for making sure that all FMS hardware and software are operational and function as a system.
5. It is important for FMS purchaser personnel to assist in system installation and start-up because of the tremendous training opportunity in working side by side with supplier personnel. True or False?
6. _____ is the single most important personnel investment in FMS.
7. All FMS training should take place during and after FMS installation. True or False?
8. The _____ are the critical focal points establishing and maintaining a healthy purchaser–supplier relationship.
9. List and explain some general guidelines that can help maintain a solid purchaser–supplier relationship and keep the project on a steady course.

19

FMS
Implementation

OBJECTIVES

After studying this unit, you will be able to:

☐ Describe the purpose and importance of acceptance testing to FMS success

☐ List and discuss the three widely used factors to measure system performance

☐ Explain the principal maintenance concerns of an automated cell or system

☐ Identify some critical factors contributing to continued system support and enhancements

INTRODUCTION

Implementation involves making all the installed components work and function as a system. It involves optimizing and fine tuning each FMS component, including computer hardware and software, to obtain peak performance. And it means making sure all personnel across all shifts are adequately trained and functioning as a team.

FMS implementation is the transition phase to decrease reliance on supplier and vendor personnel and increase reliance on in-house personnel and expertise. Supplier and OEM vendor personnel will begin to withdraw as their equipment comes on-line and is accepted. Although the FMS equipment will be under warranty for some time, making the system perform to expectation and productivity levels is now squarely up to the system purchaser.

ACCEPTANCE TESTING

Acceptance testing is making sure the entire system and all its processing modules run and perform to expected performance and specification levels as a complete and operational system (Figure 19-1). Generally, the bigger and more complex a cell or system is the longer the customer demand for a continuous, uninterrupted acceptance test.

The purpose of acceptance testing is to have the system purchaser conditionally accept the system. Usually this means that in order to be formally accepted the system must run and flawlessly perform for a specified number of continuous shifts. The number of continuous operating shifts can vary widely, but usually the required number is from nine to fifteen. Acceptance testing is the time for fine tuning and honing the system to peak performance.

Figure 19-1   Completed and operational FMS. (Courtesy of LTV Aircraft Products Group)

Acceptance testing, although scheduled for a predefined period of time as part of the project master plan, is generally agreed on jointly between system purchaser and supplier during the planning phase. This joint agreement should establish the acceptance testing ground rules, along with when complete system testing should begin and how long the complete system testing should last. The purchaser project manager approves final performance acceptance of all equipment modules and final system acceptance. It is the system supplier's responsibility to demonstrate final system acceptance, involving total system operation with all components to meet or exceed specifications as set forth and established in the functional requirements document.

To make sure acceptance testing occurs on time and on schedule and in turn keep the entire project on time and on schedule, many preliminary activities and tasks must have previously taken place and be entirely completed. These are:

1. Each equipment module has been singularly run off and individually performs to functional specifications.
2. All computer hardware and software are installed, operational, and functioning properly.
3. Purchaser personnel have been trained and learning has occurred through demonstrable and observable competencies.
4. The complete system has been started and tested for operation and functionality as an entire system.
5. Work orders and other executable commands for the entire system have been tried and tested during the initial start-up and debug phase.

Additionally, the system purchaser and supplier will be checking, recording, and verifying other aspects of system acceptance for specification compliance. Items to be checked and verified during acceptance testing include:

1. Productivity and part throughput levels
2. Equipment accuracy, functionality, and utilization levels
3. Tooling and fixturing availability and deployment
4. Processing equipment downtime for maintenance and service
5. Material-handling performance reliability and dependability
6. Part quality, accuracy, and repeatability
7. Incoming and outgoing system work flow
8. Flume system capacity and dependability
9. Computer software feature execution and verification

Software is the most error-prone component of an FMS. This is due to the size, complexity, and interdependency of the various application software modules. Consequently, full software feature testing under a variety of production situations during the alloted acceptance testing time may be difficult. However, system warranty helps protect the system purchaser as software errors or bugs are discovered.

It has been common practice on many system installations and acceptance tests for supplier personnel to actually run and perform the acceptance tests and user personnel to observe. This is true in many situations because the installed system has not been formally turned over to the user until after the system has been accepted. However, a better idea might be to have the new user perform the acceptance test with supplier personnel observing, assisting, and guiding. Such a change from the norm is important because:

1. After final acceptance, the supplier team leaves. And even though the user team has been trained, they have had little time functioning as a complete team with the new system and with a complete system in actual production.
2. Adults learn by doing, and having the user team run the acceptance test instead of the supplier team intensifies and lengthens the actual training time. This places and keeps the user in the student's role and the supplier in the instructor's role.

PERFORMANCE GOALS AND EXPECTATIONS

Any automated cell or system has estimated performance goals and expectations. Until the system is in complete operation and full production over an extended period of time, no one knows for sure whether the performance goals were really attained, exceeded, or unattainable.

System performance goals and expectations are established early in the planning phase and begin to quickly materialize during system acceptance testing and implementation. Many factors influence or help determine whether the system is performing to anticipated levels of performance. However, the following factors are widely used to effectively measure system performance:

1. Workpiece scheduling and throughput
2. Fixture flow
3. Tool flow

Workpiece scheduling and throughput are measures of how many parts are processed and how long it takes similar and unique parts to be processed through the system over a given period of time. For a manufacturing system with a high-variety, low-quantity production schedule, flexibility is imperative. An important aspect of flexibility is multiple parts per pallet of different workpiece types. Many FMSs can support a multiplicity of unique parts per pallet (Figure 19-2). Workpiece scheduling should vary enough to be flexible and fully test system flexibility during implementation.

Throughput, or workload flow, begins at the storage or workpiece input area. Workpieces are loaded on the designated fixtured pallets and transferred via the material-handling system to the various workstations for part processing. When all processing is complete, the pallet is sent to the load–unload station for part removal. Finished parts exit the system, and unfinished parts are placed in queue

Figure 19-2   Many cells and systems can support a multiplicity of unique parts per pallet. (Courtesy of Cincinnati Milacron)

until they are scheduled to enter the system for additional processing. How quickly various workpieces enter and exit the system measures part throughput and workload flow.

Fixture flow (Figure 19-3) is a measure of how quickly fixtures can be determined, built, and married to the pallet to accommodate workpiece requirements. Fixtured pallets are transported to the pallet storage area where they wait until required at the load–unload station. Delays in fixture determination, assembly, marriage to the pallet, delivery, or part loading can affect system performance levels and expectations.

Timely tool flow must also be maintained in order for the cell or system to obtain anticipated performance levels. Tool groups or kits required for a workload are defined by the planning aids. When the plan is entered into the FMS operational data base, the list of tools to build for each kit is made available to the tool preset operator through a computer terminal. The proper tool assemblies, including backup redundant tools for the required workload, must be built, gaged to length and diameter, transported to the proper machine tool, and loaded in the tool matrix. Used tools are loaded onto the tool pallet and returned to the tool room for disposition.

Although the majority of both tool and fixture flow occurs while the FMS is running and processing other parts, maintaining this simultaneous activity is critical to avoiding delays and attaining expected performance goals. The total system is only as efficient as its most inefficient flow.

Figure 19-3   How quickly fixtures can be assembled, married to the pallets, and brought into the FMS is an important factor to FMS success. (Courtesy of Remington Arms)

MAINTENANCE CONCERNS

During installation and start-up, the automated cell or system will have a high degree of "infant mortality." That is, many individual component and system failures will occur. However, from installation through acceptance testing, and even through the warranty period, the system supplier is ultimately responsible for system maintenance. The system purchaser assumes system maintenance responsibilities only after the warranty period has expired.

Maintenance concerns cover three general areas:

1. *Hardware:* all processing and material-handling equipment and computers
2. *Software:* computer operating software and FMS application software
3. *Facilities:* heating, ventilation, air conditioning, electrics, plumbing, and other building utilities

Facility maintenance concerns have been and continue to be the responsibility of the system user. The user is responsible for constructing, modernizing, and maintaining the facility and all related utilities that support the cell or system.

Software maintenance involves maintaining and updating both the operating system software (supplied by the host computer vendor) and the FMS application software (supplied by the system supplier or subcontracted software supplier or, in some cases, by the FMS user).

Operating system software can best be handled after the warranty period through a renewable maintenance contract with the computer system vendor. Application software, if supplied by the system supplier or through a subcontracted software vendor, should also be handled through a renewable maintenance contract. Without a maintenance contract, the temptation for the user to attempt software coding changes or enhancement is great. Yielding to the temptation and making minor FMS software changes can have major disastrous effects on software functionality and system up time. Changes to vendor-supplied application software should only be done by the system supplier. User application software changes can void the warranty and create other software problems undetected until a later date. Maintenance contracts help protect the system supplier, contracted vendors, and ultimately the system user.

If the FMS application software is being written by the system user software team, maintaining the software is still a maintenance concern because of personnel turnover, proper training, and adequate documentation. Just because in-house knowledge and expertise existed to initially write, debug, and apply complex FMS code, it does not necessarily mean that that level of knowledge will continue to be maintained. An astute succession plan should be put in place to properly train backup resources as, inevitably, employee turnover occurs. And adequate documentation must be developed and maintained to support the life blood of the FMS.

Hardware maintenance covers a variety of complex equipment electrically, mechanically, and hydraulically. Computer hardware maintenance is generally covered by a maintenance contract either with the computer vendor or through the system supplier to the computer vendor. As the computer is the heart of an FMS, a purchaser cannot afford to be without adequate computer hardware coverage.

Processing equipment maintenance, once the warranty expires, is generally in the hands of the system user. The best insurance for good maintenance is good training (Figure 19-4). And, with such a variety of processing equipment along with increasing technical complexity, maintaining the FMS equipment modules is a principal concern. Increased FMS utilization levels, as we have already studied, place an increased burden on all equipment in a system. Consequently, the ability of the system user to provide prompt in-house service and maintenance is a high priority.

Common practice has dictated that when something goes wrong either before or during the warranty period the supplier is notified to fix the problem. Overreliance on supplier expertise to diagnose and resolve system problems during the installation, acceptance testing, and warranty periods allows for purchaser maintenance skills to deteriorate. A more applicable philosophy would be to make the FMS user the front-line maintenance team during the warranty period, with supplier personnel as backup resources. This is a similar practice to that described under Acceptance Testing in that it forces more internal self-reliance and support and maximizes user team training, teamwork, and learning effectiveness.

Preventive maintenance of various equipment is the system purchaser's responsibility. Highly utilized equipment must have regular preventive maintenance in order to maintain adequate equipment uptime and performance reliability. The old adage "if it ain't broke, don't fix it" may have worked in older and less

Figure 19-4 Good training is an important consideration to continued maintenance and support. (Courtesy of Cincinnati Milacron)

advanced manufacturing industries, but with automated cells and systems, preventive maintenance is as much a part of system activity as the actual production work. FMS software conveniently allows for taking machines out of production by rescheduling and rerouting work to other equipment.

The important aspects about preventive maintenance, or PM as it is commonly referred to, are diligence and consistency. Although unpopular, painstaking effort must be applied to preventive maintenance procedures and a consistent and regular schedule adopted and rigorously followed.

Now that the cell or system is installed and operational, it is each shift's system manager or team leader's responsibility to make sure all maintenance guidelines and procedures are followed, along with preventive maintenance compliance.

CONTINUED SUPPORT

Once the automated cell or system is installed, acceptance tested, and fully implemented (Figure 19-5), the project is essentially completed and the project teams are disbanded. The new system's user teams covering first, second, and third shift are now operating at full capacity with the FMS system managers or team leaders for each shift now managing and directing system activities.

The supplier and purchaser together have controlled the project, and now the user must maintain or exceed system performance goals and protect the investment through continued support. Continued support means developing a high

Figure 19-5   Completely installed and fully implemented FMS. (Courtesy of Cincinnati Milacron)

degree of internal self-reliance and discipline to provide ongoing assistance, maintenance, and human resource support. Although the culmination of much of what has already been discussed, the elements of continued support include:

1.   Maintenance Concerns

Although the system supplier and its OEM vendors are available, strong internal ability to handle the wide variety of equipment maintenance problems is essential to system continued support. Well-orchestrated and continuous training is the best way to make sure system maintenance concerns are covered.

2.   Job Rotation

One of the biggest problems facing a new cell or system user is how to keep the team members motivated following exuberant and aggressive installation and implementation activity. Now that the system is completely operational, efforts must be taken to overcome worker complacency and the anticlimatic routinization of daily operation. Job rotation is one avenue available to help keep people challenged, promote growth opportunity and motivation, and provide a means for continuous training. Rotating team assignments can be an effective way of securing continued support for an automated system.

3.   Increased Team Authority

In some cases, automated system teams have become self-managing and self-auditing. As more advanced equipment becomes available on the shop floor and as systems become more integrated, lower-level employees will have more authority about how they do their jobs. This pushes ownership and support responsibilities as far down in the organization as possible. Although this could speed up decision making and create greater job satisfaction, it could also threaten middle managers and first-line supervisors.

4.   Maintain Close Contact with System Suppliers

Just because a new system is fully implemented and operational does not necessarily mean that this is as far as a company should go with automation plans and improvements. By maintaining close contact with the system supplier and other vendors and promoting internal innovation, additional improvements, enhancements, and ideas may be discovered and incorporated in the system for even greater productivity and performance improvements. By staying current with technology and seeking additional modernization changes and enhancements, system stagnation can help be prevented and long-term survivability, productivity and profitability increased.

5.   Increasing Workload and System Expansion

Long before the system is installed or implemented, an increased workload and/or system expansion may be planned. Such long-term phase II types of plans may be justified and implemented based on the success of the current or phase I system. Continued expansion and increased workload can help ensure long-term system success, particularly in the event of rapid product change and obsolescence.

6.   Safety

Nothing can shake the foundation of automation success like an accident with personal injury that should have been prevented. Worker safety must be first and foremost in system planning and implementation in order to protect employees and continued system success.

7.   Integrate with Other Computer Systems

Integration with other computer systems (CAD/CAM, MRP) is one of the surest ways to help ensure the continued long-term success of a cell or system. Linking an FMS to other existing computer systems makes the automated cell a contributing member of a larger and expanded system. This extends a system's automation capabilities and is a vital step toward CIM.

8.   Operate System Conservatively

To minimize the possibility of system downtime, machines in the system need to be operated at conservative feeds and speeds (in some cases 80 percent of optimum) and with light depths of cut. Running conservative feeds and speeds helps minimize continued internal support (manual intervention) while holding down tool inventory costs.

SUMMARY

1. The purpose of acceptance testing is to have the system purchaser conditionally accept the system.

2. The purchaser project manager approves final performance acceptance of all equipment and final system acceptance.

3. It is the system supplier's responsibility to demonstrate final system acceptance.

4. Software is the most error-prone component of an FMS.

5. The three most widely used factors to effectively measure system performance are: workpiece scheduling and throughput, fixture flow, and tool flow.

6. The total system is only as efficient as its most inefficient flow.

7. The system supplier is ultimately responsible for system maintenance from installation and acceptance testing through the warranty period.

8. The system purchaser assumes system maintenance responsibilities after the warranty period has expired.

9. Maintenance concerns cover three general areas: hardware, software, and facilities.

10. The best insurance for good maintenance is good training.

11. Continued support means developing a high degree of internal self-reliance and discipline to provide ongoing assistance, maintenance, and human resource support.

REVIEW QUESTIONS

1. The purpose of acceptance testing is to have the system _____ conditionally accept the system.

2. Items to be checked and verified during acceptance testing include:
   (a) Productivity and part throughput levels
   (b) Part quality, accuracy, and repeatability
   (c) Flume system capacity and dependability
   (d) Only (a) and (b)
   (e) (a), (b), and (c)

3. Software is not the most error-prone component of an FMS. True or False?

4. System performance goals and expectations are:
   (a) Not developed until the system is installed
   (b) Established early in the planning phase
   (c) Developed only by the system user
   (d) Always able to be attained

5. Fixture flow is a measure of how long it takes similar and unique parts to be processed through the system over a given period of time. True or False?

6. The system _____ is ultimately responsible for system maintenance from installation and acceptance testing through the warranty period.

7. Changes to vendor supplier application software should only be done by the system _____ .

8. Computer hardware maintenance is generally covered by a maintenance contract either directly with the computer vendor or through the system supplier. True or False?

9. The best insurance for good maintenance is good _____ .

10. Preventive maintenance for various equipment is the system _____ responsibility.

11. Increasing FMS team authority and responsibility can speed up decision making but could threaten supervision. True or False?

12. Explain the importance of integrating a cell or FMS with other computer systems for continued success.

A

Avoiding Problems and Pitfalls

Installation and implementation of a cell or FMS is an undertaking of considerable risk, investment, and importance for most companies. Success or failure to a large degree depends on how well the upfront planning and ongoing communication efforts by both the purchaser and supplier team have been executed. The following axioms are major points of concern for anyone associated with automated manufacturing. They apply to top and middle management, as well as the engineers, technicians, and operating personnel who must make the cell or system perform.

1. Sound planning is the single most important factor to achieving success in automated manufacturing.
2. Detailed part analysis is the single most important factor in deciding how big a cell or system will be and what it will do.
3. Planning works best when it deals with current issues and the foreseeable future and when it outlines immediate actions and benchmarked results.
4. Management must be committed through personal involvement to champion the cause and to provide the necessary communication, support, direction, and leadership to keep the project tied together, moving, and on schedule.
5. Highly successful automation projects are more likely to occur in companies that plan from the top down and implement from the bottom up.
6. The need for cells and systems in manufacturing is not as important as the need for flexibility in manufacturing.
7. Traditional justification techniques based on ROI and direct labor cost reduc-

tion cannot accommodate the cost avoidance and productivity enhancements available with flexible automation systems.

8. Many FMS limitations originate from unrealistic expectations as to what FMS is and what it can do and misunderstanding about the need for flexibility in manufacturing.

9. Implementation efforts should begin with low-cost, low-tech improvements such as reducing setup time and improving quality and material flow patterns before implementing high-cost, high-tech improvements.

10. Detailed and well-prepared technical specifications by the cell or FMS user project team are the best way to minimize the risk of the supplier or vendors underestimating the task.

11. The project master plan is *the* controlling document for the entire automation effort. It establishes commitments, responsibilities, deadlines and activities. It must be respected and rigorously followed.

12. Tool management is one of the most difficult but vital aspects of FMS to regulate and control.

13. Teamwork, communication, and partnership are the key ingredients between cell or system user and supplier to make the project succeed.

14. Organizing and staffing for automation require changes in the way a company organizes its plant and manages its people. How an FMS should be staffed depends on how well it will be organized and what it will do.

15. The best insurance for good system maintenance and support is good training.

16. FMS requires more technically astute people because of its increased complexity and computerized dependencies. Develop internal self-reliance and discipline through carefully developed and orchestrated training to provide ongoing system support and maintenance and reduce vendor dependency.

17. Cultural change problems are more difficult to solve than technical problems.

18. Cells and systems are industry and application specific, and their very flexibility requires that each cell or system be configured differently. Talk to vendors, review information, and get educated before developing specifications.

19. Establish a unified manufacturing strategy by first developing an overall business plan that makes quality improvements, cost reduction, and shortened lead time major goals.

20. Make design for manufacturability and continuous improvement a way of life; map out business goals and strategies, benchmark milestones, and target results.

21. Cells and systems often work better than their support facilities, which sometimes means that the cell or FMS must operate at less than its maximum throughput capabilities.

22. Training is as important and integral to automation as software is to a computer.

23. Be sure the level of flexibility being planned is what is needed. Excessive flexibility creates confusion and leads to higher costs, difficult and unwieldy operating procedures, and management problems.

24. The fundamental requirement with either a cell or FMS is to target the part family or families to be produced in order to scope out and determine the size, type, and configuration of the cell of system.

25. A phased FMS installation based on a beginning cellular approach to manufacturing gives people learning time to utilize and maximize the system's productive capabilities; however, it can be a drawback if it is spread over too long a period of time.

26. The cell or system purchase–supplier relationship is a marriage for the duration of the project, and the success of that relationship depends upon how well it developed and matured during the early planning stage.

B

General Safety Rules for Automated Equipment

1. Wear safety glasses and safety shoes at all times.
2. Do not wear neckties, long sleeves, wristwatches, rings, gloves, and the like, around operating equipment.
3. Maintain good housekeeping. Make sure the area around automated equipment is well lighted, dry, and free from clutter and obstructions.
4. Pay attention to all warning lights, beacons, and audible signals.
5. Hand tools and accessories should be kept off the equipment and all its moving units. Do not use machine elements as a workbench.
6. Avoid bumping any processing equipment or control units.
7. Never perform grinding operations near automated equipment. Abrasive dust will cause undue wear, inaccuracies, and possible failure of affected parts.
8. Never place hands near a revolving spindle. Keep hands away from moving equipment units.
9. Perform setup, tool, and fixture adjustments, as well as workpiece checks, with the spindle stopped.
10. Make sure parts and fixtures are securely clamped before beginning machining.
11. Use only properly sharpened tools.
12. Do not use compressed air to blow chips from the part or to clean equipment surfaces, cabinets, controls, or the general work area.
13. When handling or lifting parts or tooling, follow company policy on correct procedures.

14. Work platforms should be sturdy and must have antislip surfaces.

15. When handling tools or changing tools by hand, use a glove or shop rag. Avoid contact with cutting edges. Do not operate equipment while still wearing gloves.

16. Exercise caution when changing tools and avoid interference with the fixture or workpiece.

17. Never manually operate any automated equipment without proper training or supervision. Consult the specific operator's manual for that particular machine and control type.

18. Never attempt to program any automated equipment without proper training or supervision. Consult the specific programming manual for that specific machine and control type.

19. Do not attempt to alter automated cell or system hardware or software without the proper authorization and approval.

20. Electrical compartment doors should be opened only for electrical and/or maintenance work. They should be opened only by experienced electricians and/or qualified service personnel.

21. Safety guards, covers, and other devices have been provided for protection. Do not operate any equipment with these devices disconnected, removed, or out of position. Operate equipment only when they are in proper operating condition and position.

EIA and AIA National Codes

G word	Explanation.
G00	Used for denoting a rapid traverse rate with point-to-point positioning.
G01	Used to describe linear interpolation blocks and reserved for contouring.
G02, G03	Used with circular interpolation.
G04	A calculated time delay during which there is no machine motion (dwell).
G05,G07	Unassigned by the EIA. May be used at the discretion of the machine tool or system builder; could also be standardized at a future date.
G06	Parabolic interpolation.
G08	Acceleration code that causes the machine, assuming the capability, to accelerate at a smooth exponential rate.
G09	Deceleration code that causes the machine, assuming the capability, to decelerate at a smooth exponential rate.
G10–G12	Normally unassigned for CNC systems. Used with some hard-wired systems to express blocks of abnormal dimensions.
G13–G16	Used to direct the control system to operate on a particular set of axes.
G17–G19	Used to identify or select a coordinate plane for such functions as circular interpolation or cutter compensation.

G20–G32	Unassigned according to EIA standards; however, may be assigned by the control system or machine tool builder.
G33–G35	A mode selected for machines equipped with thread-cutting capabilities and generally referring to lathes. G33 is used when a constant lead is desired, G34 is used when a constantly increasing lead is required, and G35 is employed to designate a constantly decreasing lead.
G36–G39	Unassigned.
G40	A command that will terminate any cutter compensation.
G41	A code associated with cutter compensation in which the cutter is on the left side of the work surface, looking in the direction of the cutter motion.
G42	A code associated with cutter compensation in which the cutter is on the right side of the work surface.
G43, G44	Used with cutter offset to adjust for the difference between the actual and programmed cutter radii or diameters. G43 refers to an inside corner, and G44 refers to an outside corner.
G45–G49	Unassigned.
G50–G59	Reserved for adaptive control.
G60–G69	Unassigned.
G70	Inch programming.
G71	Metric programming.
G72	Three-dimensional circular interpolation (CW).
G73	Three-dimensional circular interpolation (CCW).
G74	Cancel multiquadrant circular interpolation.
G75	Multiquadrant circular interpolation.
G76–G79	Unassigned.
G80	Cancel cycle.
G81	Drill, or spot drill, cycle.
G82	Drill with a dwell.
G83	Intermittent, or deep hole, drilling.
G84	Tapping cycle.
G85–G89	Boring cycles.
G90	Absolute input. Input data is to be in absolute dimensional form.
G91	Incremental input. Input data is to be in incremental form.
G92	Preload registers to desired values, e.g., preload axis position registers.
G93	Inverse time feed-rate.
G94	Inches (millimeters) per minute feed-rate.
G95	Inches (millimeters) per revolution feed-rate.
G97	Spindle speed in revolutions per minute.
G98, G99	Unassigned.

MISCELLANEOUS FUNCTIONS

M word	Explanation.
M00	Program stop. Operator must cycle start in order to continue with the remainder of the program.
M01	Optional stop. Acted on only when the operator has previously signaled for this command by pushing a button. When the control system senses the M01 code, the machine will automatically stop.
M02	End of program. Stops the machine after completion of all commands in the block. May include rewinding of tape.
M03	Start spindle rotation in a clockwise direction.
M04	Start spindle rotation in a counterclockwise direction.
M05	Spindle stop.
M06	Command to execute the change of a tool (or tools) manually or automatically.
M07	Turn coolant on (flood).
M08	Turn coolant on (mist).
M09	Coolant off.
M10, M11	Automatic clamping of the machine slides, workpiece, fixture, spindle, and the like. M11 is an unclamping code.
M12	An inhibiting code to synchronize multiple sets of axes, such as a four-axis lathe having two independently operated heads or slides.
M13	Combines simultaneous clockwise spindle motion and coolant on.
M14	Combines simultaneous counterclockwise spindle motion and coolant on.
M15, M16	Rapid traverse or feed motion in either the $+$ (M15) or $-$ (M16) direction.
M17, M18	Unassigned.
M19	Oriented spindle stop. Spindle stop at a predetermined angular position.
M20–M29	Unassigned.
M30	End-of-tape command. Will rewind the tape and also automatically transfer to a second tape reader if incorporated in the control system.
M31	A command known as interlock bypass for temporarily circumventing a normally provided interlock.
M32–M39	Unassigned.
M40–M46	Used to signal gear changes if required at the machine; otherwise, unassigned.
M47	Continues program execution from the start of the program unless inhibited by an interlock signal.
M48	Cancel M49.

M49	A function that deactivates a manual spindle or feed override and returns to the programmed value.
M50–M57	Unassigned.
M59	A function that holds the rpm constant at its value when M59 is initiated.
M60–M99	Unassigned.

OTHER ADDRESS CHARACTERS

Address Character	Explanation
A	Angular dimension about the X axis.
B	Angular dimension about the Y axis.
C	Angular dimension about the Z axis.
D	Can be used either to express an angular dimension around a special axis, a third feed function, or tool offset.
E	Also used for angular dimension around a special axis or a second feed function.
H	Unassigned.
I, J, K	Used with circular interpolation.
L	Not used.
O	Used in place of the customary sequence number word address N.
P	A third rapid traverse code or tertiary motion dimension parallel to the X axis.
Q	Second rapid traverse code or tertiary motion dimension parallel to the Y axis.
R	First rapid traverse code or tertiary motion dimension parallel to the Z axis or the radius for constant surface speed calculation.
U	Secondary motion dimension parallel to the X axis.
V	Secondary motion dimension parallel to the Y axis.
W	Secondary motion dimension parallel to the Z axis.

D

Metric Conversion Chart

CONVERSION CHART (Based on 25.4 mm = 1 in.) Inches into Millimeters

Inches		M M	Inches		M M	Inches	M M	Inches	M M	Inches	M M
1/64	0.0156	0.3969	49/64	0.7656	19.4469	34	863.600	82	2082.80	130	3302.00
1/32	0.0313	0.7937	25/32	0.7813	19.8437	35	889.000	83	2108.20	131	3327.40
3/64	0.0469	1.1906	51/64	0.7969	20.2406	36	914.400	84	2133.60	132	3352.80
1/16	0.0625	1.5875	13/16	0.8125	20.6375	37	939.800	85	2159.00	133	3378.20
5/64	0.0781	1.9844	53/64	0.8281	21.0344	38	965.200	86	2184.40	134	3403.60
3/32	0.0938	2.3812	27/32	0.8438	21.4312	39	990.600	87	2209.80	135	3429.00
7/64	0.1094	2.7781	55/64	0.8594	21.8281	40	1016.00	88	2235.20	136	3454.40
1/8	0.1250	3.1750	7/8	0.8750	22.2250	41	1041.40	89	2260.60	137	3479.80
9/64	0.1406	3.5719	57/64	0.8906	22.6219	42	1066.80	90	2286.00	138	3505.20
5/32	0.1563	3.9687	29/32	0.9063	23.0187	43	1092.20	91	2311.40	139	3530.60
11/64	0.1719	4.3656	59/64	0.9219	23.4156	44	1117.60	92	2336.80	140	3556.00
3/16	0.1875	4.7625	15/16	0.9375	23.8125	45	1143.00	93	2362.20	141	3581.40
13/64	0.2031	5.1594	61/64	0.9531	24.2094	46	1168.40	94	2387.60	142	3606.80
7/32	0.2188	5.5562	31/32	0.9688	24.6062	47	1193.80	95	2413.00	143	3632.20
15/64	0.2344	5.9531	63/64	0.9844	25.0031	48	1219.20	96	2438.40	144	3657.60
1/4	0.2500	6.3500	1		25.4000	49	1244.60	97	2463.80	145	3683.00
17/64	0.2656	6.7469	2		50.800	50	1270.00	98	2489.20	146	3708.40
9/32	0.2813	7.1437	3		76.200	51	1295.40	99	2514.60	147	3733.80
19/64	0.2969	7.5406	4		101.600	52	1320.80	100	2540.00	148	3759.20
5/16	0.3125	7.9375	5		127.000	53	1346.20	101	2565.40	149	3784.60
21/64	0.3281	8.3344	6		152.400	54	1371.60	102	2590.80	150	3810.00
11/32	0.3438	8.7312	7		177.800	55	1397.00	103	2616.20	151	3835.40
23/64	0.3594	9.1281	8		203.200	56	1422.00	104	2641.60	152	3860.80

Fraction	Decimal in.	mm	No.	mm	No.	mm	No.	mm	No.	mm
3/8	0.3750	9.5250	9	228.600	57	1447.80	105	2667.00	153	3886.20
25/64	0.3906	9.9219	10	254.000	58	1473.20	106	2692.40	154	3911.60
13/32	0.4063	10.3187	11	279.400	59	1498.60	107	2717.80	155	3937.00
27/64	0.4219	10.7156	12	304.800	60	1524.00	108	2743.20	156	3962.40
7/16	0.4375	11.1125	13	330.200	61	1549.40	109	2768.60	157	3987.80
29/64	0.4531	11.5094	14	355.600	62	1574.80	110	2794.00	158	4013.20
15/32	0.4688	11.9062	15	381.000	63	1600.20	111	2819.40	159	4038.60
31/64	0.4844	12.3031	16	406.400	64	1625.60	112	2844.80	160	4064.00
1/2	0.5000	12.7000	17	431.800	65	1651.00	113	2870.20	161	4089.40
33/64	0.5156	13.0969	18	457.200	66	1676.40	114	2895.60	162	4114.80
17/32	0.5313	13.4937	19	482.600	67	1701.80	115	2921.00	163	4140.20
35/64	0.5469	13.8906	20	508.000	68	1727.20	116	2946.40	164	4165.60
9/16	0.5625	14.2875	21	533.400	69	1752.60	117	2971.80	165	4191.00
37/64	0.5781	14.6844	22	558.800	70	1778.00	118	2997.20	166	4216.40
19/32	0.5938	15.0812	23	584.200	71	1803.40	119	3022.60	167	4241.80
39/64	0.6094	15.4781	24	609.600	72	1828.80	120	3048.00	168	4267.20
5/8	0.6250	15.8750	25	635.000	73	1854.20	121	3073.40	169	4292.60
41/64	0.6406	16.2719	26	660.400	74	1879.60	122	3098.80	170	4318.00
21/32	0.6563	16.6687	27	685.800	75	1905.00	123	3124.20	171	4343.40
43/64	0.6719	17.0656	28	711.200	76	1930.40	124	3149.60	172	4368.80
11/16	0.6875	17.4625	29	736.600	77	1955.80	125	3175.00	173	4394.20
45/64	0.7031	17.8594	30	762.000	78	1981.20	126	3200.40	174	4419.60
23/32	0.7188	18.2562	31	787.400	79	2006.60	127	3225.80	175	4445.00
47/64	0.7344	18.6531	32	812.800	80	2032.00	128	3251.20		
3/4	0.7500	19.0500	33	838.200	81	2057.40	129	3276.60		

0.001 in. = 0.0254 mm   0.001 mm = 0.0004 in

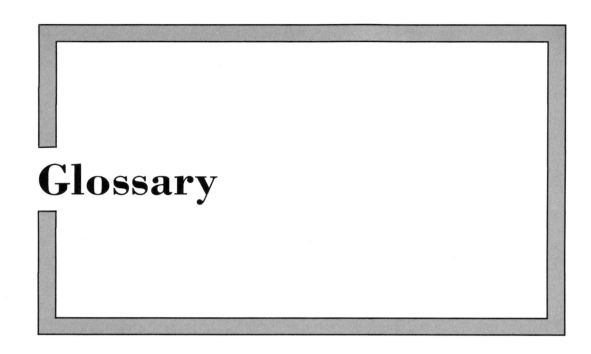

Glossary

A axis — Circular motion about the X axis (sometimes referred to as alpha axis).

Absolute coordinate — Measurement of existing position from the zero point.

Absolute system — Numerically controlled system in which all input and feedback positional dimensions are given with respect to a common reference point.

Accandec — Acceleration and deceleration. Pertains to feed-rate motion and provides smooth starts and stops on NC equipment when feed-rate changes are required.

Access time — The time interval between the instant data are called out from the storage unit and the instant they are delivered to the processing unit (read time); the time interval between the instant data are requested to be stored and the instant at which storage is completed (write time).

Accuracy — Maximum error permitted under specified circumstances.

AD-APT — An Air Force-adapted NC contouring program based on the APT language with limited vocabulary.

Adaptive control — Automated process of altering feeds and speeds to optimum condition by sensing cutting action.

Address — A label, name, or number identifying a location or device.

AGV — Automated guided vehicle.

Algorithm — Computational procedure for solving a problem. When properly applied, an algorithm always produces a solution to a problem.

Alphanumeric characters — Loosely used to apply to alphabetical characters and decimal digits A through Z and 0 through 9, respectively.

Analog — Use of a physical quantity, such as voltage, with amplitude representative of another physical quantity, such as distance.

Analog computer — A computer that makes analogies between numbers and directly measurable quantities, such as voltage and rotations.

Analysis — The investigation of a problem by some consistent, systematic procedure.

Analyst — A person skilled in the definition of and the development of techniques for the solving of a problem, especially those techniques for solutions on a computer.

ANSI — American National Standards Institute. An organization that acts as a national clearinghouse and coordinator for voluntary standards in the United States. Formerly the USA Standards Institute (USASI) and American Standards Association (ASA).

APT — Automatic programmed tools. A general-processor, NC computer-assist programming language for multiaxis contour programming.

Architecture — The way in which the parts of a system fit and communicate with one another.

ASCII — American Standard Code for Information Interchange. One of the main paper tape formats.

ASRS — Automatic storage and retrieval system.

Assembler (assembly) — The software routine that translates source or other language programs into machine-readable language or object code.

Asynchronous — Implies the ability to wait until told to go.

Automation — The implementation of processes by automatic means.

Auxiliary function — Noncoordinate and cutter movement of a machine tool that requires programming.

Axis — Primary direction that determines the relative movement of the workpiece or tool.

Axis inhibit — NC feature providing the capability to withold machine tool command information.

Axis inversion — Same as mirror image; an NC feature providing plus and minus value reversal, thereby permitting machining of the left-hand part from the right-hand part information, and vice versa.

Batch processing — Technique by which items to be processed must be coded and collected into groups prior to processing.

BCD — Binary coded decimal. Representation of a number by groups of four binary digits for each decimal in the number.

Binary — (a) Condition or situation having only two possibilities. (b) Number representation system with a radix of two.

Binary digit — A digit in a binary scale of notation representing 0 (zero) or 1 (one); a binary 1 can be represented in a preallocated place by a steady voltage, pulse, or hole, whereas a binary 0 is represented by the absence of a steady voltage, pulse, or hole.

Bit — A binary digit or its representation.

Block — A set of things, such as digits, characters, or words, handled as a unit.

Block diagram — Graphic representation showing the logical sequence in which data are processed.

Branch — Technique used to transfer control from one sequence of a program to another.

Bridge — Equipment that connects two or more LANs that use the same protocol, allowing communication between devices on separate LANs.

Broadcast message — Message from one user sent to all workstations and devices on a LAN.

Buffer storage — Electronic circuit that forms a temporary store for data or information signals; buffer-stored information can be instantly transferred to active storage for command execution.

Bug — Term used to denote a mistake in a computer program or malfunction in a computer hardware component.

Byte — A 16-bit word that can be assembled from 2 bytes, each of 8 bits, or from 4 bytes of 4 bits each.

CAD — Computer-aided design.

Call — Branch to a subroutine.

CAM — Computer-aided manufacturing.

CAM-I — Computer-aided Manufacturing International. A new replacement of the APT long-range planning committee.

Cartesian coordinates — Provides the capability of positioning a point with respect to a set of axes at right angles to each other.

Central processing unit (CPU) — That portion of the hardware of a computing system containing the control unit, arthrimitic unit, and internal storage unit.

Character — Set of letters, decimal digits, and special signs (+ , − , %, and so on) that may be combined to express information.

Chip — A piece of silicon or other crystalline material that is treated so that it can contain and support an integrated circuit.

Circular interpolation — A contour machine with this facility cuts an arc of a circle from one block on a control tape, generally in a single plane (XY, YZ, or ZX).

CL data — Output from a processor with information regarding cutter location.

Closed-loop system — System in which the output, or some result of the output, is fed back for comparison with the input for the purpose of reducing the difference.

CNC — Computer numerical control.

Code — System through which characters and their associations represent information.

Coding — Process of translating problem logic represented by a flow chart into computer instructions and data.

Command — A pulse, signal, or set of signals initiating a performance.

Compatible — One control system is compatible with another when they both accept the same tapes and perform the same task.

Compile — To generate a machine language program from a computer program written in a high-level source language.

Compiler — Computer program used to translate high-level source language programs into machine language programs suitable for execution on a particular computing system.

Computer — Calculating device that processes data represented by a combination of discrete data (in digital computers) or continuous data (in analog computers).

Console — That part of a computer used for communication between the computer operator or maintenance engineer and the computer.

Console printer — Auxiliary printer used in several computer systems for communications between the computer operator and the computing system.

Controller — Device between the host and terminals that relays information between them.

Coordinate system — Series of intersecting planes or planes and cylinders (usually three) that forms a reference system in which the position of a point, line, circle, or other geometric feature can be specified.

CPU — Central processing unit. The section of a computer that includes instruction execution or logic.

CRT — Cathode ray tube. Electronic vacuum tube with a screen for visual display of output information in graphical or alphanumeric form. Display is produced by means of proportionally deflected electron beams.

Cycle — Sequence of operations performed in a predetermined manner, and frequently repeated in a complete machine tool movement.

Data — Information, generally in the form of words, symbols, numbers, letters, characters, digits, and the like.

Data base — Comprehensive data file containing information in a format applicable to a user's needs and available when needed.

Debug — To rectify faults and remove mistakes in a control system or program; troubleshoot.

Decimal code — Code in which each allowable position has one out of ten possible states.

Decimal digit — One of the symbols 0 through 9 when used in numbering in base 10.

Diagnostic — Information on a printout to show how the input part program has deviated from the normal rules.

Digit — Character used in a numbering system.

Digital — Adjective describing a discrete state of being such as on or off; a combination of these forms a specific value.

Digital/analog converter — Device providing an analog quantity corresponding to a digital value.

Disk pack — The vertical stacking of a series of magnetic disks in a removable self-contained unit.

DNC — Direct numerical control.

Documentation — Printed materials, such as manuals, that provide usage information for a manufactured product.

Downtime — Time during which equipment is inoperable because of faults.

Dynamic memory — Memory device that stores data as a charge or a capacitive component and therefore needs refreshing regularly in order to retain its content.

Edit — To change or modify the main program or format of data.

EIA — Electronics Industries Association. One of the main paper tape formats.

File — Organized collection of related data. For example, the entire set of inventory master data records make up the Inventory Master File.

Firmware — Programs or control instructions that are not changeable (by the user) and that are held in read-only memory (ROM) or another permanent memory device.

Floating point number — Expression comprising an integer and/or fractional part stored in a computer in floating point form.

FMS — Flexible manufacturing system.

Format — General makeup of items, including arrangement and location of all information.

Gateway — Hardware and software that permit LANs using different protocols to communicate with each other.

General processor — Computer program for converting geometric input data into cutter path data of a general nature; general processor information that generally requires postprocessing.

Group technology — Grouping or arranging of parts through design characteristics or features and/or manufacturing processes.

Hard copy — Printed and visually readable information on paper.

Hardware — Equipment and component parts of a controller or computer, such as integrated circuits, and transistors.

Hard-wired — Fixed interconnection of logic circuits on a circuit board to cause a predetermined sequence of events to occur.

Highway — Major route or path for data or information signals.

Hub — Center of a star topology network, often a file server, that houses the network software and direct communications within the network. It can also act as the gateway to another LAN.

Inhibit — Prevention of action, acceptance, or reading of data by applying a signal and stopping the input.

Input — Transfer of external information into a control system.

Input media — Form of input information, such as punched cards or magnetic tape.

Intelligent terminal — Terminal that can be programmed.

Interchange station — A wait station where a tool on a tool-changing machine waits to transfer either to the spindle or the tool matrix drum station.

Interface — Relationship between software modules that are usually in the same mode.

Intermediate transfer arm — Tool-exchanging device that removes and places tools in the spindle, interchange station, and tool drum.

Interpolation — Function involving the generation of control or data points between the given coordinate positions.

ISO — International Standards Organization. One of the main paper tape formats (same as ASCII).

Letter address — Alphabetic character (X, Y, Z, T, G, and so on) that precedes the numeric part of an NC code or coordinate and directs its registration to the designated location in the system.

Line — Physical path that provides direct communication among a number of stations.

Local area network (LAN) — Data communications network spanning a limited geographical area, such as an office, an entire building, or industrial park. It provides communication between computers and peripherals.

Machine code — Code obeyed by a computer or microprocessor system with no need for further translation (normally written in binary code).

Machine language — Basic language of a computer. Programs written in machine language equivalent of a source program.

Machining center — NC machine tool, capable of changing its own cutting tools, that performs successive machining operations, such as drilling, milling, tapping, and boring.

Macro — Collection of instructions that can be stored and recalled on a repetitive basis (also called a subroutine).

Magnetic tape — Plastic tape coated with a thin layer of magnetic material that is activated in rows of seven to nine spots across the tape to represent a character.

Magnetic tape unit — Device used to read and write data in the form of magnetic spots on reels of tape coated with a magnetizable material.

Main frame — The part of the computer that contains the arithmetic unit, internal storage unit, and control functions. Same as the central processing unit.

Management information system (MIS). — An all-inclusive system designed to provide instant data to management for effective and efficient business operation.

Manual data input (MDI) — Mode of operator control providing direct insertion into the control system.

Manual part programming — Manuscript preparation in machine control language and format to define the command sequence for an NC machine (G, X, Y, Z, and so on, words).

Manuscript — Printed or written copy of machine code information identical to that acted on by the machine.

Media — Communications channels over which LANs operate. These include coaxial cable, twisted-pair telephone wiring, and fiber-optic cabling.

Memory — General term applied to internal storage of data, programs, or other information needed by a computer.

Modal — Active information that is retained and in effect until replaced by new information (stays in effect until changed or canceled).

Multiplexer — Combination of hardware and software that allows simultaneous transmission and reception of two or more data streams on a single channel.

NC — Numerical control; control of machine tools by numbers through command instructions in coded form.

Network architecture — Set of rules, standards, or recommendations through which various computer hardware, operating systems, and applications software function together.

Node — Point in a network where service is provided or used or where communications channels are interconnected.

Numerical control system — System in which actions are controlled by the direct insertion of programmed numerical data.

Object program — Computer program in the internal language of a particular computer; the machine language equivalent of a source program.

Off-line operation — Preparation of tapes, workpieces, and the like, away from the processing or utilization function.

Offset — Axial displacement of the tool that is the difference between the actual and programmed tool length.

On line — Peripheral devices operating under the direct control of the central processing unit.

On-line operation — Where tape punching or workpiece setup is done at the function utilization site; also used to denote direct computer hookup for instantaneous response.

Operating system — Software that controls the execution of computer programs and that may provide scheduling, input–output control, compilation, data management, debugging, storage assignment, accounting, and other similar functions.

Optimize — Rearrangement of instructions to obtain the minimum number of data transfers or part and cutter movements.

Output — Data transferred from a computer's internal storage unit to storage or an output device.

Packet — Unit of data to be routed from a source node to a destination node.

Part programmer — Specialist who prepares part programs for the operation of numerically controlled machine tools.

PLC — Programmable logic controller.

Plotter — Electromechanical device that draws a plot or traces cutter path information from NC input.

Port — Point at which data can enter or leave a network, such as the serial or parallel ports in the back of most PCs.

Postprocessor — Computer program that converts all coordinate data and machine motion commands from the general processor into a machine control readable format.

Processor — Program that translates a source program into object language.

Program — Systematic arrangement of instructions to be executed by a control or computer in order to perform a specific function.

Protocol — Set of messages with specific formats and rules for exchanging the messages.

Quadrant — One of the four sections into which a plane is divided by 90 degrees, intersecting coordinate axes in that plane.

Random — Not in logical order, sequence, or arrangement, but having the flexibility to select from any location and in any order.

Random access memory (RAM) — Memory device that can be written and read under program control and that is often used as a scratchpad memory by the control logic or by the user for his or her data.

Rapid traverse — Maximum feed movement between cutting positions.

Reader — Device that senses bits of information on punched cards, punched tape, or magnetic tape.

Read-only memory (ROM) — Memory device containing information that is fixed and can only be read and not changed.

Register — Special store location used for specific logic or control functions for temporary storage of data.

Repeatability — Maximum difference within a series of results when the same demand is continuously applied under identical conditions.

Routine — Set of functionally related instructions that directs the computer to carry out a desired operation. A subdivision of a program.

Server — Module or set of modules that performs a well-defined service, such as remote file access or gateway communication, on behalf of another module.

Significant digit — Digit used to preserve specific accuracy or placement that must be kept.

Simulate — To represent the functioning of one system by another; that is, to represent a physical system by the execution of a computer program or to represent a biological system by a mathematical model.

Simulator — Software development tool that executes a program almost as the final system would, but slowly, so as to produce a program trace.

Software — Computer programs and instructional documentation used to assist in programming, operating, and maintaining computers and manufacturing systems.

Source program — Computer program written in a symbolic programming language (for example, assembly language program, FORTRAN program, COBOL program). A translator is used to convert the source program into an object program that can be executed on a computer. Contrasted with object program.

SPC — Statistical process control.

Statement — Agreed-upon arrangement of words and/or data accepted by a system to command a particular computer function.

Storage — Memory capacity into which information can be held and extracted as required at a later time.

Subroutine — Sequence of computer programming statements or instructions that performs frequently required operations (also called a macro).

Terminal — Unit in a network, such as a modem or telephone, at which data can be either sent or received.

Tool length compensation — Manual input process that allows the programmer to program all tools as if they were of equal length.

Tool offset — Tool position correction, parallel to a controlled axis, normally used to compensate for tool wear.

Topology — Physical arrangement and relationship of interconnected nodes and lines in a network.

Turnkey system — Name applied to a system sold by a manufacturer who has total responsibility for building, installing, programming, and testing the system for conformance to customer specifications.

USASCII — United States of America Standard Code for Information Interchange (full name for ASCII).

Verify — To check, usually with an automatic machine, one typing or recording of data against another in order to discover punching or transmission errors in the data transcription.

Volatile Storage — Storage medium in which data cannot be retained without continuous power dissipation.

Word — Defined arrangement of characters and digits that conveys one instruction or piece of information.

X axis — Axis of horizontal motion and parallel to the work-holding surface.

Y axis — Axis of motion that is perpendicular to the X and Z axes.

Z axis — Axis of motion parallel to the machine spindle.

Zero shift — Numerically controlled machine tool capability that enables the zero point on an axis to be changed or shifted within a specified range; the control does not retain permanent zero information.

Selected Bibliography

Flexible Manufacturing Systems. Boston: Yankee Group, 1985. A market study of the users, vendors, and types of flexible manufacturing systems. Lists the benefits of a flexible system. Gives a detailed description of the planning and implementation of an FMS.

Flexible Manufacturing Systems Handbook. Park Ridge, NJ: Noyes Publications, 1984. Provides a methodical approach to the acquisition, applications, requirements, and design of FMS. Includes overview and summary sections, as well as detailed technical material on specific steps that must be taken. Provides a glossary of terms that are unique or have special meaning in the FMS field.

FMS Report. Bedford, England: IFS Publications, 1984. A revised study of worldwide FMS installations, conducted by Ingersoll Engineers. Besides basic background information on economic justification, requirements, and the like, a number of FMS applications, including illustrated plant layouts and specifications, are discussed.

HARRINGTON, JOSEPH. *Computer Integrated Manufacturing*. New York: Industrial Press, 1973. An early comprehensive work that covers the philosophy, economics, and techniques of computer-integrated manufacturing and forecasts the greater things yet to come. Breaks down complex manufacturing systems into basic elements for easier analysis.

HARTLEY, JOHN. *FMS at Work*. Bedford, England: IFS Publications Ltd., 1984. Explains why FMS is advantageous; describes the essential elements and tells how some companies are putting the principles to good use. Also suggests the potential for improved systems in the future and what might happen to employment.

HOLLAND. JOHN R., *Flexible Manufacturing Systems,* 1984. Dearborn, MI: Society of Manufacturing Engineers, 1984. A compendium of articles that attempts to define what is really meant by FMS and how this concept differs from conventional manufacturing thought. Deals with technological, business, social, and economic implications.

KOREN, YORAM. *Computer Control of Manufacturing Systems*. New York: McGraw-Hill, 1983. Provides an introduction to the theory and applications of control in the manufacturing area. Presents concepts of computer control as applied to stand-alone manufacturing systems, such as machine tools and industrial robots, and provides a useful approach to their implementations.

RANKY, PAUL. *Design and Operation of FMS*. Bedford, England: IFS Publications Ltd., 1983. Written for practicing engineers as well as for researchers and managers to help analyze, design , implement, and run FMS. This comprehensive text of FMS practice demonstrates the state of the art and the expected trends.

REMBOLD, ULRICH, ARMBRUSTER, KARL, AND ULZMANN, WOLFGANG. *Interface Technology for Computer Controlled Manufacturing Processes*. New York: Marcel Dekker, 1983. Presents a thorough discussion of the hardware and software aspects of computer interfacing and the interactions of these two technologies. Helpful for students and application engineers of process control and instrumentation.

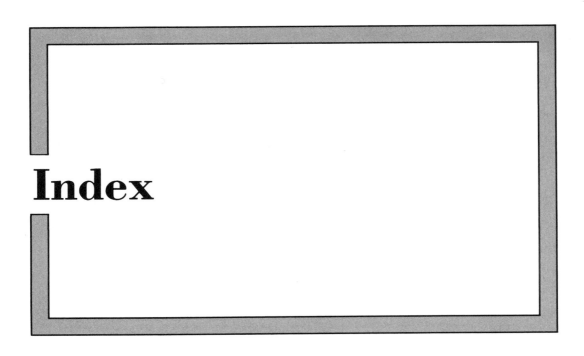

Index